Joint FAO/WHO Food Standards Programme
CODEX ALIMENTARIUS COMMISSION

CODEX ALIMENTARIUS
VOLUME TWO
PESTICIDE RESIDUES IN FOOD

FOOD AND AGRICULTURE ORGANIZATION
OF THE UNITED NATIONS
WORLD HEALTH ORGANIZATION
Rome, 1993

M-83
ISBN 92-5-103271-8

INTRODUCTION

The Maximum Residue Limits (MRLs) and other recommendations appearing in this Volume were adopted by the Codex Alimentarius Commission acting on the advice of its Committee on Pesticide Residues. They are consistent with the recommendations of the Joint FAO/WHO Meetings on Pesticide Residues (JMPR).

The JMPR is composed of two panels of experts[1] who serve in their personal capacity only and who do not represent or otherwise present the views of their governments or organizational affiliation. These experts meet jointly to evaluate selected pesticide residues on the basis of available data obtained from various sources, including governments, industry and academia. Where appropriate, they establish "Acceptable Daily Intakes" (ADIs) and recommend MRLs for residues of pesticides on foods and animal feeds.

Acceptance of Codex MRLs for pesticide residues by Member Nations and Associate Members of FAO and/or WHO, proceeds as described in Section 1, Volume 1 (1992), under "General Principles of the Codex Alimentarius".

---

[1] *The FAO Expert Panel on Pesticide Residues in Food and the Environment and the WHO Expert Panel on Pesticide Residues.*

CONTENTS

SECTION 1

LIST OF CODEX
MAXIMUM RESIDUE LIMITS
FOR PESTICIDES

LIST OF CODEX MAXIMUM RESIDUE LIMITS
FOR PESTICIDES

INTRODUCTION

The present volume contains all Maximum Residue Limits for pesticides adopted by the Commission up to and including its 19th Session. Further adoptions by the Commission will be published as an Addendum to Volume 2. The pesticides in this Section are listed in numerical order of the Codex code number. The missing numbers represent pesticides for which (a) Codex maximum limits are being elaborated, (b) for which previously established Codex MRLs have been withdrawn, or (c) for which only "guideline levels" have been recommended by the Joint Meeting on Pesticide Residues pending toxicological clearance of the pesticide and its residues.

BASIS FOR ESTABLISHMENT OF CODEX MAXIMUM RESIDUE LIMITS AND CODEX EXTRANEOUS RESIDUE LIMITS

Codex "Maximum Residue Limits" (MRLs) are recommended on the basis of appropriate residue data obtained mainly from supervised trials. The residue data thus obtained reflect registered or approved usage of the pesticide in accordance with "good agricultural practices". These may vary considerably from region to region owing to differences in local pest control requirements which are due to a variety of reasons. Consequently, residues in food, particularly at a point close to harvest may also vary. In establishing Codex MRLs, these variations in residues due to differences in "good agricultural practices" are taken into consideration, as far as possible on the basis of available data.

As Codex MRLs cover a wide spectrum of use patterns and "good agricultural practices" and need to reflect residue levels closely following harvest, they may occasionally be higher than the levels of residues found in national surveillance activities. This may be especially so with easily degradable pesticides and when analysis is carried out at a point in the distribution chain far removed from the last application of the pesticide.

Codex MRLs are established only where there is supporting evidence concerning the safety to humans of the resulting residues as determined by the Joint FAO/WHO Meeting on Pesticide Residues and this means that Codex Maximum Residue Limits represent residue levels which are toxicologically acceptable.

Another type of Codex Maximum Limit, the Codex "Extraneous Residue Limit" (ERL) which covers residues arising from environmental contamination or uses of pesticides other than agricultural uses. It is mainly based on residue data obtained from national food control or monitoring activities. Codex ERLs need to cover widely varying residue levels in food reflecting differing situations in respect of contamination of food by environmental and persistent pesticide residues. For this reason, Codex ERLs cannot always reflect strictly the actual local residue situation existing in given countries or regions. Codex ERLs represent acceptable residue levels which are intended to facilitate international trade in food while protecting the health of the consumer. They are established only when there is supporting evidence concerning the safety to humans of the residues as determined by the Joint FAO/WHO Meeting on Pesticide Residues.

CODEX MAXIMUM RESIDUE LIMITS AND CONSUMER PROTECTION: DETERMINATION OF TOTAL DAILY INTAKE OF PESTICIDE RESIDUES

The primary purpose of setting maximum limits for pesticide residues in food and in some cases, in animal feeds, is to protect the health of the consumer. Codex MRLs and ERLs serve that primary purpose as they help to ensure that only the minimum amount of pesticide is applied to food consistent with real pest control needs. Codex Maximum Residue Limits are based on residue data from supervised trials and not directly derived from Acceptable Daily Intakes (ADIs), which are a quantitative expression of acceptable daily amounts of residue which persons may ingest on a long term basis and which are established on the basis of appropriate toxicological data mainly from animal studies.

The acceptability of Codex Maximum Residue Limits is judged on the basis of a comparison of the acceptable daily intake with estimated daily intakes, as determined on the basis of suitable intake studies. Intake data from such studies, compared with acceptable daily intakes, help in determining the safety of foods in respect of pesticide residues. Guidelines for predicting Dietary Intakes of Pesticide Residues have been prepared under the joint sponsorship of UNEP, FAO and WHO[1].

CODEX MAXIMUM RESIDUE LIMITS FOR MILK AND MILK PRODUCTS

Codex Maximum Residue Limits for fat-soluble pesticide residues in milk and milk products are expressed on a whole product basis.

For a "milk product" with a fat content less than 2%, the MRL applied should be half those specified for milk. The MRL for "milk products" with a fat content of 2% or more should be 25 times the maximum residue limit specified for milk, _expressed on a fat basis_.

Fat soluble pesticide residues to which the above general provision applies, are indicated by means of the letter "F" in conjunction with the MRL specified for milk.

CODEX MAXIMUM RESIDUE LIMITS FOR PROCESSED FOODS

As a rule, Codex MRLs and ERLs are established for raw agricultural commodities, However, where it is considered necessary for consumer protection and facilitation of trade. MRLs and ERLs are also established for certain processed foods on a case-by-case basis, taking into consideration information on the influence of processing on residues.

---

[1]   (Ref. Guidelines for Predicting Dietary Intake of Pesticide Residues, Joint UNEP/FAO/WHO, World Health Organization, Geneva 1989).

EXPLANATORY NOTES

The foods listed in the columns marked "commodity" shall not contain more than the maximum amount (in mg/kg) stated in the columns marked "Maximum Residue Limit", of the pesticide residue (defined in each individual case in the definition of residue) at (a) the point of entry into a country or (b) at the point of entry into trade channels within a country. This maximum limit shall not be exceeded at any time thereafter.

The Codex maximum residue limits (MRLs) apply to the residue content of the final sample representative of the lot and of the portion of commodities which is analyzed. (See Section 4 of this Volume).

Dates of Estimation of Acceptable Daily Intakes (ADIs)

ADIs - The year of estimation or confirmation is shown in parenthesis.

TADIs - The period from the year of estimation or most recent extension to the year by which data needed for the estimation of a full ADI are required, is shown in parenthesis.

Notes on the MRLs

(*) (following MRLs) : At or about the limit of determination.

E (following MRLs) : Extraneous Residue Limit (ERL).

F (following MRLs for milk) : The residue is fat soluble and MRLs for milk products are derived as explained earlier in this Section.

(Fat) (following MRLs for meat) : The MRLs apply to the fat of meat.

Po (following MRLs) : The MRL accommodates post-harvest treatment of the commodity.

PoP (following MRLs for processed foods) : The MRL accommodates post-harvest treatment of the primary food commodity.

T (following MRLs) : The MRL is temporary, irrespective of the status of the ADI, until required information has been provided and evaluated.

V (following MRLs for products of animal origin) : The MRL accommodates veterinary uses.

ALPHABETIC INDEX OF PESTICIDE CHEMICALS FOR WHICH MAXIMUM RESIDUE LIMITS

HAVE BEEN RECOMMENDED OR ARE UNDER CONSIDERATION

NUMERICAL INDEX OF PESTICIDE CHEMICALS FOR WHICH MAXIMUM RESIDUE LIMITS

HAVE BEEN RECOMMENDED OR ARE UNDER CONSIDERATION

1 ALDRIN AND DIELDRIN	51 METHIDATHION	101 PIRIMICARB	151 DIMETHIPIN
2 AZINPHOS-METHYL	52 METHYL BROMIDE	102 MALEIC HYDRAZIDE	152 FLUCYTHRINATE
3 BINAPACRYL	53 MEVINPHOS	103 PHOSMET	153 PYRAZOPHOS
4 BROMOPHOS	54 MONOCROTOPHOS	104 DAMINOZIDE	154 THIODICARB
5 BROMOPHOS-ETHYL	55 OMETHOATE	105 DITHIOCARBAMATES	155 BENALAXYL
6 CAPTAFOL	56 ORTHO-PHENYLPHENOL	106 ETHEPHON	156 CLOFENTEZINE
7 CAPTAN	57 PARAQUAT	107 ETHIOFENCARB	157 CYFLUTHRIN
8 CARBARYL	58 PARATHION	108 ETHYLENETHIOUREA (ETU)	158 GLYPHOSATE
9 CARBON DISULPHIDE	59 PARATHION-METHYL	109 FENBUTATIN OXIDE	159 VINCLOZOLIN
10 CARBON TETRACHLORIDE	60 PHOSALONE	110 IMAZALIL	160 PROPICONAZOLE
11 CARBOPHENOTHION	61 PHOSPHAMIDON	111 IPRODIONE	161 PACLOBUTRAZOL
12 CHLORDANE	62 PIPERONYL BUTOXIDE	112 PHORATE	162 TOLYLFLUANID
13 CHLORDIMEFORM	63 PYRETHRINS	113 PROPARGITE	163 ANILAZINE
14 CHLORFENVINPHOS	64 QUINTOZENE	114 GUAZATINE	164 DEMETON-S-METHYLSULPHONE
15 CHLORMEQUAT	65 THIABENDAZOLE	115 TECNAZENE	165 FLUSILAZOLE
16 CHLOROBENZILATE	66 TRICHLORFON	116 TRIFORINE	166 OXYDEMETON-METHYL
17 CHLORPYRIFOS	67 CYHEXATIN	117 ALDICARB	167 TERBUFOS
18 COUMAPHOS	68 AZINPHOS-ETHYL	118 CYPERMETHRIN	168 TRIADIMENOL
19 CRUFOMATE	69 BENOMYL	119 FENVALERATE	169 CYROMAZINE
20 2,4-D	70 BROMOPROPYLATE	120 PERMETHRIN	170 HEXACONAZOLE
21 DDT	71 CAMPHECHLOR	121 2,4,5-T	171 PROFENOFOS
22 DIAZINON	72 CARBENDAZIM	122 AMITRAZ	172 BENTAZONE
23 1,2-DIBROMOETHANE	73 DEMETON-S-METHYL	123 ETRIMFOS	173 BUPROFEZIN
24 1,2-DICHLOROETHANE	74 DISULFOTON	124 MECARBAM	174 CADUSAFOS
25 DICHLORVOS	75 PROPOXUR	125 METHACRIFOS	175 GLUFOSINATE-AMMONIUM
26 DICOFOL	76 THIOMETON	126 OXAMYL	176 HEXYTHIAZOX
27 DIMETHOATE	77 THIOPHANATE-METHYL	127 PHENOTHRIN	
28 DIOXATHION	78 VAMIDOTHION	128 PHENTHOATE	
29 DIPHENYL	79 AMITROLE	129 AZOCYCLOTIN	
30 DIPHENYLAMINE	80 CHINOMETHIONAT	130 DIFLUBENZURON	
31 DIQUAT	81 CHLOROTHALONIL	131 ISOFENPHOS	
32 ENDOSULFAN	82 DICHLOFLUANID	132 METHIOCARB	
33 ENDRIN	83 DICLORAN	133 TRIADIMEFON	
34 ETHION	84 DODINE	134 AMINOCARB	
35 ETHOXYQUIN	85 FENAMIPHOS	135 DELTAMETHRIN	
36 FENCHLORPHOS	86 PIRIMIPHOS-METHYL	136 PROCYMIDONE	
37 FENITROTHION	87 DINOCAP	137 BENDIOCARB	
38 FENSULFOTHION	88 LEPTOPHOS	138 METALAXYL	
39 FENTHION	89 SEC-BUTYLAMINE	139 BUTOCARBOXIM	
40 FENTIN	90 CHLORPYRIFOS-METHYL	140 NITROFEN	
41 FOLPET	91 CYANOFENPHOS	141 PHOXIM	
42 FORMOTHION	92 DEMETON	142 PROCHLORAZ	
43 HEPTACHLOR	93 BIORESMETHRIN	143 TRIAZOPHOS	
44 HEXACHLOROBENZENE	94 METHOMYL	144 BITERTANOL	
45 HYDROGEN CYANIDE	95 ACEPHATE	145 CARBOSULFAN	
46 HYDROGEN PHOSPHIDE	96 CARBOFURAN	146 CYHALOTHRIN	
47 INORGANIC BROMIDE	97 CARTAP	147 METHOPRENE	
48 LINDANE	98 DIALIFOS	148 PROPAMOCARB	
49 MALATHION	99 EDIFENPHOS	149 ETHOPROPHOS	
50 MANCOZEB	100 METHAMIDOPHOS	150 PROPYLENETHIOUREA (PTU)	

LIST OF CODEX MAXIMUM RESIDUE LIMITS
FOR PESTICIDES

<u>References</u>

(1)   **Code Number and Name of Commodity** - Reference to the Codex Classification of Foods and Animal Feeds [see Section 2].

(2)   **JMPR** - Year of the JMPR evaluation or subsequent review.

(3)   **CCPR** - Number of CCPR session and paragraph of the related CCPR report (i.e. 23.105).

(1) **ALDRIN AND DIELDRIN**

JMPR   65, 66, 67, 68, 69, 70, 74, 75, 77, 90
ADI   0.0001 mg/kg body weight | (Confirmed 1977)
Residue   Sum of HHDN and HEOD (fat-soluble).
Note   23rd CCPR (1991) agreed with the recommendation of the 1990 JMPR
to convert existing CXLs to TERLs, pending monitoring data (23.71).

Commodity code No.	Name	MRL (mg/kg)		
VS 0621	Asparagus	0.1		E T
VB 0400	Broccoli	0.1		E T
VB 0402	Brussels sprouts	0.1		E T
VB 0041	Cabbages, Head	0.1		E T
VR 0577	Carrot	0.1		E
VB 0404	Cauliflower	0.1		E T
GC 0080	Cereal grains	0.02		E
VC 0424	Cucumber	0.1		E T
VO 0440	Egg plant	0.1		E T
PE 0112	Eggs	0.1		E
A02 0001	Fruits	0.05		E T
VR 0583	Horseradish	0.1		E T
VL 0482	Lettuce, Head	0.1		E
MM 0095	Meat	0.2	(fat)	E
ML 0106	Milks	0.006	F	E
VA 0385	Onion, Bulb	0.1		E T
VR 0588	Parsnip	0.1		E T
VO 0051	Peppers	0.1		E T
VO 0445	Peppers, Sweet	0.1		E T
VR 0589	Potato	0.1		E T
VR 0494	Radish	0.1		E T
VL 0494	Radish leaves	0.1		E T

(2) AZINPHOS-METHYL

```
JMPR    65, 68, 72, 73, 74, 91
  ADI   0.005 mg/kg body weight  | (1991)
Residue   Azinphos-methyl.
```

Commodity code No.	Name	MRL (mg/kg)	
AL 1021	Alfalfa forage (green)	2	
TN 0660	Almonds	0.2	
FS 0240	Apricot	2	1/
VB 0400	Broccoli	1	
VB 0402	Brussels sprouts	1	1/
VS 0624	Celery	2	1/
GC 0080	Cereal grains	0.2	
FC 0001	Citrus fruits	2	1/
SO 0691	Cotton seed	0.2	
A02 0002	Fruits (except..)	1	1/ 2/
FB 0269	Grapes	4	
FI 0341	Kiwifruit	4	
VC 0046	Melons, except Watermelon	2	
AL 0528	Pea vines (green)	2	1/
FS 0247	Peach	4	
VR 0589	Potato	0.2	
VD 0541	Soya bean (dry)	0.2	
AL 1265	Soya bean forage (green)	2 fresh wt	
SO 0702	Sunflower seed	0.2	1/
A01 0002	Vegetables (except..)	0.5	1/ 2/

1/ JMPR 1991 recommended to withdraw the MRL

2/ (Except as otherwise listed)

(7) CAPTAN

```
JMPR    63, 65, 69, 73, 74, 77, 78, 80, 82, 84, 86, 87, 90
 ADI    0.1 mg/kg body weight  | (1984 and confirmed 1990)
Residue   Captan.
```

	Commodity code No.	Name	MRL (mg/kg)		
FP	0226	Apple	25	T	1/
FB	0020	Blueberries	20	T	1/
FC	0001	Citrus fruits	15	T	2/
DF	0269	Dried grapes	5	T	2/
FS	0247	Peach	15	T	1/
FP	0230	Pear	25	T	1/
FB	0275	Strawberry	20	T	1/
VO	0448	Tomato	15	T	1/

1/ JMPR 1990 had proposed that temporary levels would remain until
   1992, pending receipt of data and GAP information.

2/ JMPR 1990 had proposed to withdraw the CXL in view of not
   expected uses.

(8) CARBARYL

JMPR   65, 66, 67, 68, 69, 70, 73, 75, 76, 77, 79, 84
  ADI   0.01 mg/kg body weight | (1973)
Residue   Carbaryl.

Commodity code No.	Name		MRL (mg/kg)	
AL 1021	Alfalfa forage (green)		100	
FP 0226	Apple		5	
FS 0240	Apricot		10	
VS 0621	Asparagus		10	
FI 0327	Banana		5	
GC 0640	Barley		5	Po
AL 1030	Bean forage (green)		100	
VR 0574	Beetroot		2	
FB 0264	Blackberries		10	
FB 0020	Blueberries		7	
VB 0041	Cabbages, Head		5	
VR 0577	Carrot		2	
MM 0812	Cattle meat		0.2	
FS 0013	Cherries		10	
FC 0001	Citrus fruits		7	
AL 1023	Clover		100	fresh wt.
VP 0526	Common bean (pods and/ or immature seeds)		5	
SO 0691	Cotton seed		1	
VD 0527	Cowpea (dry)		1	
FB 0265	Cranberry		7	
VC 0424	Cucumber		3	
FB 0266	Dewberries		10	
VO 0440	Egg plant		5	
PE 0112	Eggs		0.5	
MM 0814	Goat meat		0.2	
FB 0269	Grapes		5	
AS 0162	Hay or fodder (dry) of grasses		100	
FI 0341	Kiwifruit		10	
VL 0053	Leafy vegetables		10	
AF 0645	Maize forage		100	fresh wt
VC 0046	Melons, except Watermelon		3	
A03 0001	Milk products		0.1	(*)
ML 0106	Milks		0.1	(*)
FS 0245	Nectarine		10	
A05 1900	Nuts (whole in shell)		10	
GC 0647	Oats		5	Po
VO 0442	Okra		10	

(8) **CARBARYL**

JMPR   65, 66, 67, 68, 69, 70, 73, 75, 76, 77, 79, 84
ADI   0.01 mg/kg body weight | (1973)
Residue   Carbaryl.

	Commodity		
code No.	N a m e	MRL (mg/kg)	
FT	0305	Olives	10	
DM	0305	Olives, processed	1	
VR	0588	Parsnip	2	
AL	0528	Pea vines (green)	100	fresh wt
FS	0247	Peach	10	
AL	0697	Peanut fodder	100	
SO	0703	Peanut, whole	2	
FP	0230	Pear	5	
VP	0063	Peas	5	
VO	0051	Peppers	5	
FS	0014	Plums (including Prunes)	10	
VR	0589	Potato	0.2	
PM	0110	Poultry meat	0.5	V
PO	0113	Poultry skin	5	V
VC	0429	Pumpkins	3	
VR	0494	Radish	2	
FB	0272	Raspberries, Red, Black	10	
GC	0649	Rice	5	Po
CM	0649	Rice, husked	5	Po
GC	0650	Rye	5	Po
MM	0822	Sheep meat	0.2	
GC	0651	Sorghum	10	Po
AF	0651	Sorghum forage (green)	100	fresh wt
VD	0541	Soya bean (dry)	1	
AL	1265	Soya bean forage (green)	100	fresh wt
VC	0431	Squash, Summer	3	
FB	0275	Strawberry	7	
VR	0596	Sugar beet	0.2	
AV	0596	Sugar beet leaves or tops	100	
VR	0497	Swede	2	
VO	1275	Sweet corn (kernels)	1	
VO	0448	Tomato	5	
TN	0085	Tree nuts	1	
GC	0654	Wheat	5	Po
CM	0654	Wheat bran, unprocessed	20	PoP
CF	1211	Wheat flour	0.2	PoP
CF	1212	Wheat wholemeal	2	PoP

(8) CARBARYL

JMPR   65, 66, 67, 68, 69, 70, 73, 75, 76, 77, 79, 84
  ADI   0.01 mg/kg body weight | (1973)
Residue Carbaryl.

Commodity code No.	Name	MRL (mg/kg)
VC 0433	Winter squash	3

(11) CARBOPHENOTHION

JMPR   72, 76, 77, 79, 80, 83
ADI   0.0005 mg/kg body weight | (1979)
Residue   Sum of carbophenothion, its sulphoxide and its sulphone, expressed as
          carbophenothion (fat-soluble).
Note   This compound may no longer be manufactured. Comments were requested
          by CL 1992/12-PR on a proposal to withdraw CXL (24.242).

Commodity code No.	Name	MRL (mg/kg)		
FP 0226	Apple	1		
FS 0240	Apricot	1		
VB 0400	Broccoli	0.5		
VB 0402	Brussels sprouts	0.5		
MM 0812	Cattle meat	1	(fat)	V
VB 0404	Cauliflower	0.5		
FC 0001	Citrus fruits	2		
ML 0106	Milks	0.004	F	V
FS 0245	Nectarine	1		
OC 0305	Olive oil, crude	0.2		
FT 0305	Olives	0.1		
FS 0247	Peach	1		
FP 0230	Pear	1		
TN 0672	Pecan	0.02	(*)	
FS 0014	Plums (including Prunes)	1		
VR 0589	Potato	0.02	(*)	
SO 0495	Rape seed	0.02	(*)	
MM 0822	Sheep meat	1	(fat)	V
VL 0502	Spinach	2		
VR 0596	Sugar beet	0.1		
TN 0678	Walnuts	0.02	(*)	

(12) CHLORDANE

JMPR   65, 67, 69, 70, 72, 74, 77, 82, 84, 86
ADI   0.0005 mg/kg body weight | (1986)
Residue   Sum of cis- and trans-chlordane, or, in the case of animal products,
sum of cis- and trans-chlordane and "oxychlordane" (fat-soluble).

Commodity code No.		Name	MRL (mg/kg)		
TN	0660	Almonds		0.02	E
OC	0691	Cotton seed oil, crude		0.05	E
PE	0112	Eggs		0.02	E
AO2	0003	Fruits and vegetables		0.02	(*) E
TN	0666	Hazelnuts		0.02	E
OC	0693	Linseed oil, crude		0.05	E
GC	0645	Maize		0.02	E
MM	0095	Meat		0.05	E (fat)
ML	0106	Milks		0.002	E, F
GC	0647	Oats		0.02	E
TN	0672	Pecan		0.02	E
PM	0110	Poultry meat		0.5	E (fat)
CM	1205	Rice, polished		0.02	E
GC	0650	Rye		0.02	E
GC	0651	Sorghum		0.02	E
OC	0541	Soya bean oil, crude		0.05	E
OR	0541	Soya bean oil, refined		0.02	E
TN	0678	Walnuts		0.02	E
GC	0654	Wheat		0.02	E

(14) CHLORFENVINPHOS

JMPR   71, 84
  ADI   0.002 mg/kg body weight | (1971)
Residue   Chlorfenvinphos, sum of E- and Z-isomers (fat soluble).

Commodity code No.	Name	MRL (mg/kg)
VB 0400	Broccoli	0.05
VB 0402	Brussels sprouts	0.05
VB 0041	Cabbages, Head	0.05
VR 0577	Carrot	0.4
VB 0404	Cauliflower	0.1
VS 0624	Celery	0.4
FC 0001	Citrus fruits	1
SO 0691	Cotton seed	0.05
VO 0440	Egg plant	0.05
VR 0583	Horseradish	0.1
VA 0384	Leek	0.05
GC 0645	Maize	0.05
MM 0095	Meat	0.2 (fat) V
ML 0107	Milk of cattle, goats and sheep	0.008 F V
VO 0450	Mushrooms	0.05
VA 0385	Onion, Bulb	0.05
SO 0697	Peanut	0.05
VR 0589	Potato	0.05
VR 0494	Radish	0.1
GC 0649	Rice	0.05
CM 1205	Rice, polished	0.05
VR 0497	Swede	0.05
VR 0508	Sweet potato	0.05
VO 0448	Tomato	0.1
VR 0506	Turnip, Garden	0.05
GC 0654	Wheat	0.05

(15) CHLORMEQUAT

JMPR   70, 72, 76, 85
  ADI   0.05 mg/kg body weight | (1972)
Residue   Chlormequat cation (usually used as the chloride).
Note   22nd CCPR decided that the compound was candidate for cancellation
    (22.358) Review was scheduled for 1994 JMPR

Commodity code No.	Name	MRL (mg/kg)	
AS 0640	Barley straw and fodder, dry	50	
DF 0269	Dried grapes	1	
FB 0269	Grapes	1	
ML 0107	Milk of cattle, goats and sheep	0.1	(*)
A03 0001	Milk products	0.1	(*)
AS 0647	Oat straw and fodder, dry	50	
GC 0647	Oats	10	
FP 0230	Pear	3	
GC 0650	Rye	5	
AS 0650	Rye straw and fodder, dry	50	
GC 0654	Wheat	5	
AS 0654	Wheat straw and fodder, dry	50	

(16) CHLOROBENZILATE

JMPR   65, 68, 72, 75, 77, 80
  ADI   0.02 mg/kg body weight | (confirmed 1980)
Residue   Chlorobenzilate.
Note   This compound may no longer be manufactured. Comments were requested
       by CL 1992/12-PR on a proposal to withdraw CXL (24.242).

Commodity code No.	Name	MRL (mg/kg)	
TN 0660	Almonds	0.2	
FP 0226	Apple	5	
FC 0001	Citrus fruits	1	
FB 0269	Grapes	2	
VC 0046	Melons, except Watermelon	1	
ML 0107	Milk of cattle, goats and sheep	0.05	(*)
FP 0230	Pear	2	
VO 0448	Tomato	0.2	
TN 0678	Walnuts	0.2	

(17) CHLORPYRIFOS

JMPR   72, 74, 75, 77, 81, 82, 83, 89(GAP information only).
ADI   0.01 mg/kg body weight | (1982)
Residue   Chlorpyrifos (fat-soluble).

Commodity code No.	Name	MRL (mg/kg)	
FP 0226	Apple	1	
VB 0041	Cabbages, Head	0.05	(*)
VR 0577	Carrot	0.5	
MM 0812	Cattle meat	2	(fat) V
VB 0404	Cauliflower	0.05	(*)
VS 0624	Celery	0.05	(*)
PM 0840	Chicken meat	0.1	(fat)
VL 0467	Chinese cabbage, type "Pe-tsai"	1	
FC 0001	Citrus fruits	0.3	
VP 0526	Common bean (pods and/ or immature seeds)	0.2	
SO 0691	Cotton seed	0.05	(*)
OC 0691	Cotton seed oil, crude	0.05	(*)
DF 0269	Dried grapes	2	
VO 0440	Egg plant	0.2	
PE 0112	Eggs	0.05	(*)
FB 0269	Grapes	1	
VL 0480	Kale	1	
FI 0341	Kiwifruit	2	
VL 0482	Lettuce, Head	0.1	
ML 0106	Milks	0.01	(*) F V
VO 0450	Mushrooms	0.05	(*)
VA 0385	Onion, Bulb	0.05	(*)
FP 0230	Pear	0.5	
VO 0051	Peppers	0.5	
VR 0589	Potato	0.05	(*)
FB 0272	Raspberries, Red, Black	0.2	
GC 0649	Rice	0.1	
MM 0822	Sheep meat	0.2	(fat) V
VR 0596	Sugar beet	0.05	(*)
VO 0448	Tomato	0.5	
PM 0848	Turkey meat	0.2	(fat) V

(20) **2,4-D**

JMPR   70, 71, 74, 75, 80, 85, 86, 87(correction to 1986)
ADI   0.3 mg/kg body weight | (1975)
Residue   2,4-D.
Note   2,4-D was found to have continued use and is held pending further
          information from manufacturer (23.320).

Commodity code No.	Name		MRL (mg/kg)
GC 0640	Barley	0.5	
FB 0264	Blackberries	0.1	
FC 0001	Citrus fruits	2	
PE 0112	Eggs	0.05	(*)
GC 0645	Maize	0.05	(*)
MM 0095	Meat	0.05	(*)
A03 0001	Milk products	0.05	(*)
ML 0106	Milks	0.05	(*)
GC 0647	Oats	0.5	
VR 0589	Potato	0.2	
FB 0272	Raspberries, Red, Black	0.1	
GC 0649	Rice	0.05	(*)
GC 0650	Rye	0.5	
GC 0651	Sorghum	0.05	(*)
FB 0019	Vaccinium berries, including Bearberry	0.1	
GC 0654	Wheat	0.5	

(21) **DDT**

JMPR   65, 66, 67, 68, 69, 78, 79, 80, 83, 84
  ADI   0.02 mg/kg body weight | (1984)
Residue Sum of p,p'-DDT, o,p'-DDT, p,p'-DDE and p,p'TDE (DDD) (fat-soluble).

Commodity code No.	Name	MRL (mg/kg)		
GC 0080	Cereal grains	0.1		E T 1/
PE 0112	Eggs	0.5		E T 1/
MM 0095	Meat	5	(fat)	E T 1/
ML 0106	Milks	0.05		F E T 1/

1/ Temporary limits pending review on the basis of GAP and monitoring
   data (23.77).

(22) **DIAZINON**

JMPR   65, 66, 67, 68, 70, 75, 79
ADI   0.002 mg/kg body weight | (1970)
Residue   Diazinon (fat-soluble).

Commodity code No.	Name	MRL (mg/kg)	
TN 0660	Almonds	0.1	
GC 0640	Barley	0.1	
FC 0001	Citrus fruits	0.7	
SO 0691	Cotton seed	0.1	
AO2 0002	Fruits (except..)	0.5	1/
TN 0666	Hazelnuts	0.1	
VL 0053	Leafy vegetables	0.7	
MM 0097	Meat of cattle, pigs & sheep	0.7	(fat) V
ML 0106	Milks	0.02	F V
OC 0305	Olive oil, crude	2	
FT 0305	Olives	2	
FS 0247	Peach	0.7	
SO 0697	Peanut	0.1	
TN 0672	Pecan	0.1	
CM 1205	Rice, polished	0.1	
SO 0699	Safflower seed	0.1	
SO 0702	Sunflower seed	0.1	
VO 0447	Sweet corn (corn-on-the-cob)	0.7	
AO1 0002	Vegetables (except..)	0.5	1/
TN 0678	Walnuts	0.1	
GC 0654	Wheat	0.1	

1/ (Except as otherwise listed).

(25) DICHLORVOS

JMPR   65, 66, 67, 69, 70, 74, 77
  ADI   0.004 mg/kg body weight | (confirmed 1977)
Residue Dichlorvos.
Note   Residues decline rapidly during storage and shipment. Codex MRLs are
        based on residues likely to be found at harvest or slaughter.

Commodity code No.	Name	MRL (mg/kg)	
SB 0715	Cacao beans	5	
GC 0080	Cereal grains	2	
SB 0716	Coffee beans	2	
PE 0112	Eggs	0.05	
A02 0001	Fruits	0.1	
MM 0814	Goat meat	0.05	
VD 0533	Lentil (dry)	2	
VL 0482	Lettuce, Head	1	
MM 0097	Meat of cattle, pigs & sheep	0.05	
ML 0106	Milks	0.02	
VO 0450	Mushrooms	0.5	
SO 0697	Peanut	2	
PM 0110	Poultry meat	0.05	
VD 0541	Soya bean (dry)	2	Po
A01 0002	Vegetables (except..)	0.5	1/

1/ (Except otherwise listed)

(26) **DICOFOL**

```
JMPR    68, 70, 74
 ADI    0.025 mg/kg body weight  | (1968)
Residue   Dicofol.
```

code No.	Commodity Name	MRL (mg/kg)	
VC 0424	Cucumber	2	
A02 0002	Fruits (except..)	5	1/
VC 0425	Gherkin	2	
DH 1100	Hops, dry	5	
FB 0275	Strawberry	1	
DT 1114	Tea, Green, Black	5	
VO 0448	Tomato	1	
A01 0002	Vegetables (except..)	5	1/

1/ (Except as otherwise listed)

(27) DIMETHOATE

JMPR   65, 66, 67, 70, 73(eval. of formothion)77, 78, 83, 84, 86, 87, 88, 90
ADI   0.01 mg/kg body weight | (1987)
Residue   Dimethoate
Note   See also (55) omethoate

Commodity code No.	Name	MRL (mg/kg)	
FP 0226	Apple	1	
FI 0327	Banana	1	Po
VR 0574	Beetroot	0.2	
VR 0577	Carrot	1	
VS 0624	Celery	1	
FS 0013	Cherries	2	
FC 0001	Citrus fruits	2	
FB 0278	Currant, Black	2	
FB 0269	Grapes	1	
DH 1100	Hops, dry	3	
VL 0480	Kale	0.5	
OR 0305	Olive oil, refined	0.05	(*)
FT 0305	Olives	1	
DM 0305	Olives, processed	0.05	(*)
VA 0385	Onion, Bulb	0.2	
FP 0230	Pear	1	
VP 0063	Peas	0.5	
VO 0051	Peppers	1	Po
VR 0589	Potato	0.05	
VL 0502	Spinach	1	
FB 0275	Strawberry	1	
VR 0596	Sugar beet	0.05	
AV 0596	Sugar beet leaves or tops	1	
VO 0448	Tomato	1	Po
VR 0506	Turnip, Garden	0.5	
VS 0469	Witloof chicory (sprouts)	0.5	

(29) DIPHENYL

```
JMPR    66, 67
  ADI   0.125 mg/kg body weight  | (1967)
Residue  Diphenyl.
```

Commodity code No.	N a m e	MRL (mg/kg)		
FC 0001	Citrus fruits	110	Po	1/

1/ 24th CCPR decided to recommend deletion of the CXL for citrus
   fruits (24.79).

(30) **DIPHENYLAMINE**

JMPR   69, 76, 79, 82, 84
  ADI   0.02 mg/kg body weight | (1984)
Residue   Diphenylamine.

Commodity code No.	Name	MRL (mg/kg)	
FP 0226	Apple	5	Po

----

(31) **DIQUAT**

JMPR   70, 72, 76, 77, 78
ADI   0.008 | mg diquat cation/kg body weight (1977).
Residue   Diquat cation.
Note   Generally available as dibromide (JMPR 1970).
        Scheduled for periodic review (24.241)

Commodity code No.	Name	MRL (mg/kg)	
GC 0640	Barley	5	
VP 0062	Beans, shelled	0.5	
SO 0691	Cotton seed	1	
OR 0691	Cotton seed oil, edible	0.1	
MO 0105	Edible offal (Mammalian)	0.05	(*)
PE 0112	Eggs	0.05	(*)
GC 0645	Maize	0.1	
MM 0095	Meat	0.05	(*)
ML 0106	Milks	0.01	(*)
VA 0385	Onion, Bulb	0.1	
VP 0064	Peas, shelled	0.1	
SO 0698	Poppy seed	5	
VR 0589	Potato	0.2	
SO 0495	Rape seed	2	
OR 0495	Rapeseed oil, edible	0.1	
GC 0649	Rice	5	
CM 0649	Rice, husked	0.2	
CM 1205	Rice, polished	0.2	
OR 0700	Sesame seed oil, edible	0.1	
GC 0651	Sorghum	2	
VR 0596	Sugar beet	0.1	
SO 0702	Sunflower seed	0.5	
OR 0702	Sunflower seed oil,edible	0.1	
A01 0002	Vegetables (except..)	0.05	(*) 1/
GC 0654	Wheat	2	
CM 0654	Wheat bran, unprocessed	5	
CF 1211	Wheat flour	0.2	
CF 1212	Wheat wholemeal	2	

1/ (Except as otherwise listed)

(32) ENDOSULFAN

JMPR   65, 67, 68, 71, 74, 75, 82, 85, 89
ADI   0.006 mg/kg body weight | (1989)
Residue   Sum of alpha- and beta- endosulfan and endosulfan sulphate (fat-soluble).

Commodity code No.	N a m e	MRL (mg/kg)	
AL 1021	Alfalfa forage (green)	1	
VR 0577	Carrot	0.2	
VS 0624	Celery	2	
FS 0013	Cherries	1	
AL 1023	Clover	1	
SO 0691	Cotton seed	1	
OC 0691	Cotton seed oil, crude	0.5	
A02 0001	Fruits	2	2/
VP 0528	Garden pea (young pods)	0.5	
VL 0480	Kale	1	
VL 0482	Lettuce, Head	1	
VL 0483	Lettuce, Leaf	1	
VA 0385	Onion, Bulb	0.2	
FS 0014	Plums (including Prunes)	1	
FP 0009	Pome fruits	1	
VR 0589	Potato	0.2	
GC 0649	Rice	0.1	
VL 0502	Spinach	2	
VR 0596	Sugar beet	0.1	
AV 0596	Sugar beet leaves or tops	1	
VR 0508	Sweet potato	0.2	
DT 1114	Tea, Green, Black	30	
AL 1028	Trefoil	1	
A01 0002	Vegetables (except..)	2	2/ 3/

2/ 22nd CCPR agreed to recommend deletion of the general CXL (22.270),
   however as the compound was on the agenda of the 1993, the
   Committee agreed to wait its review (23.248).

3/ (Except as otherwise listed)

(33) ENDRIN

```
JMPR    65, 70, 74, 75
 ADI    0.0002 mg/kg body weight  | (1970)
```

Residue   Sum of endrin and delta-keto-endrin (fat-soluble).

Note   22nd CCPR agreed CXLs would be mantained until sufficient monitoring
          data are available for the proposal of ERLs to replace MRLs. (22.357)

Commodity code No.		Name	MRL (mg/kg)		
FP	0226	Apple	0.02	(*)	1/
GC	0640	Barley	0.02	(*)	1/
SO	0691	Cotton seed	0.1		1/
OC	0691	Cotton seed oil, crude	0.1		1/
OR	0691	Cotton seed oil, edible	0.02	(*)	1/
PE	0112	Eggs	0.2		1/
MM	0095	Meat	0.1	E (fat)	
ML	0106	Milks -	0.0008	E F	
PM	0110	Poultry meat	1	(fat)	1/
CM	0649	Rice, husked	0.02	(*)	1/
CM	1205	Rice, polished	0.02	(*)	1/
GC	0651	Sorghum	0.02	(*)	
VO	0447	Sweet corn (corn-on-the-cob)	0.02	(*)	1/
GC	0654	Wheat	0.02	(*)	1/

1/ JMPR 1990 had no evidence that MRLs for these compounds had any
   relationship to current registered uses, and therefore recommended
   that they should be converted to TERLs, pending the receipt of
   further information.

(34) **ETHION**

JMPR   68, 69, 70, 72, 75, 82, 83, 85, 86, 89, 90
ADI   0.002 mg/kg body weight | (1990)
Residue   Ethion (fat-soluble).
Note   As the TMDI for this compound exceeds ADI, the Committee felt
          the need to review the GAP of ethion (23.93).

Commodity code No.		Name	MRL (mg/kg)	
TN	0660	Almonds	0.1	(*)
FP	0226	Apple	2	
FS	0240	Apricot	0.1	(*)
MM	0812	Cattle meat	2.5	(fat) V
MO	0812	Cattle, Edible offal of	1	
FS	0013	Cherries	0.1	(*)
TN	0664	Chestnuts	0.1	(*)
FC	0001	Citrus fruits	2	
VP	0526	Common bean (pods and/ or immature seeds)	2	
SO	0691	Cotton seed	0.5	
VC	0424	Cucumber	0.5	
VO	0440	Egg plant	1	
PE	0112	Eggs	0.2	(*)
VA	0381	Garlic	1	
MM	0814	Goat meat	0.2	(*) (fat)
MO	0814	Goat, Edible offal of	0.2	(*)
FB	0269	Grapes	2	
TN	0666	Hazelnuts	0.1	(*)
MM	0816	Horse meat	0.2	(*) (fat)
MO	0816	Horse, Edible offal of	0.2	(*)
GC	0645	Maize	0.05	(*)
VC	0046	Melons, except Watermelon	2	
ML	0106	Milks	0.02	F V
FS	0245	Nectarine	1	
VA	0385	Onion, Bulb	1	
FS	0247	Peach	1	
FP	0230	Pear	2	
TN	0672	Pecan	0.1	(*)
VO	0051	Peppers	1	
VO	0445	Peppers, Sweet	1	
MM	0818	Pig meat	0.2	(*) (fat)
MO	0818	Pig, Edible offal of	0.2	(*)
FS	0014	Plums (including Prunes)	2	
PM	0110	Poultry meat	0.2	(*)(fat)
PO	0111	Poultry, Edible offal of	0.2	(*)

(34) **ETHION**

JMPR   68, 69, 70, 72, 75, 82, 83, 85, 86, 89, 90
ADI   0.002 mg/kg body weight | (1990)
Residue   Ethion (fat-soluble).
Note   As the TMDI for this compound exceeds ADI, the Committee felt
          the need to review the GAP of ethion (23.93).

Commodity code No.	Name	MRL (mg/kg)	
MM 0822	Sheep meat	0.2	(*)(fat)
MO 0822	Sheep, Edible offal of	0.2	(*)
VC 0431	Squash, Summer	0.5	
FB 0275	Strawberry	2	
DT 1114	Tea, Green, Black	5	
VO 0448	Tomato	2	
TN 0678	Walnuts	0.1	(*)
VC 0433	Winter squash	0.5	

(35) ETHOXYQUIN

JMPR   69
  ADI   0.06 mg/kg body weight | (1969)
Residue   Ethoxyquin.
Note   22nd CCPR agreed that additional information was required to support
        CXLs (22.358).

Commodity code No.	Name	MRL (mg/kg)	
FP 0226	Apple	3	Po
FP 0230	Pear	3	Po

----

(37) **FENITROTHION**

JMPR   69, 74, 76, 77, 79, 82, 83, 84, 86, 87, 88, 89
ADI   0.005 mg/kg body weight | (1988)
Residue   Fenitrothion (fat-soluble).

code No.		Commodity Name	MRL (mg/kg)	
FP	0226	Apple	0.5	
VB	0041	Cabbages, Head	0.5	
SB	0715	Cacao beans	0.1	
VB	0404	Cauliflower	0.1	
GC	0080	Cereal grains	10	Po
FS	0013	Cherries	0.5	
FC	0001	Citrus fruits	2	
VC	0424	Cucumber	0.05	(*)
VO	0440	Egg plant	0.1	
FB	0269	Grapes	0.5	
VA	0384	Leek	0.2	
VL	0482	Lettuce, Head	0.5	
MM	0095	Meat	0.05	(*) E (fat)
ML	0106	Milks	0.002	(*) E,F
VA	0385	Onion, Bulb	0.05	(*)
FS	0247	Peach	1	
FP	0230	Pear	0.5	
VP	0063	Peas	0.5	
VO	0051	Peppers	0.1	
VR	0589	Potato	0.05	(*)
VR	0494	Radish	0.2	
CM	1206	Rice bran, unprocessed	20	PoP
CM	1205	Rice, polished	1	PoP
VD	0541	Soya bean (dry)	0.1	
FB	0275	Strawberry	0.5	
DT	1114	Tea, Green, Black	0.5	
VO	0448	Tomato	0.5	
CF	0654	Wheat bran, processed	2	PoP
CM	0654	Wheat bran, unprocessed	20	PoP
CF	1211	Wheat flour	2	PoP
CF	1212	Wheat wholemeal	5	PoP
CP	1211	White bread	0.2	PoP

(38) FENSULFOTHION

JMPR   72, 82, 83
  ADI   0.0003 mg/kg body weight | (confirmed 1982)
Residue   Sum of fensulfothion, its oxygen analogue and their sulphones,
           expressed as fensulfothion.

Commodity code No.		Name	MRL (mg/kg)	
FI	0327	Banana	0.02	(*)
MM	0812	Cattle meat	0.02	(*)(fat)
MO	0812	Cattle, Edible offal of	0.02	(*)
MM	0814	Goat meat	0.02	(*)(fat)
MO	0814	Goat, Edible offal of	0.02	(*)
GC	0645	Maize	0.1	
VA	0385	Onion, Bulb	0.1	
SO	0697	Peanut	0.05	(*)
FI	0353	Pineapple	0.05	(*)
GC	0656	Popcorn	0.1	
VR	0589	Potato	0.1	
MM	0822	Sheep meat	0.02	(*)(fat)
MO	0822	Sheep, Edible offal of	0.02	(*)
VR	0596	Sugar beet	0.1	
VR	0497	Swede	0.1	
VO	0448	Tomato	0.1	

(39) **FENTHION**

JMPR   71, 75, 77, 78, 79, 80, 83, 89
ADI   0.001 mg/kg body weight | (1980)
Residue   Sum of fenthion, its oxygen analogue and their sulphoxides and
           sulphones, expressed as fenthion (fat-soluble).
Note   Scheduled for periodic review (24.241).

Commodity code No.		Name	MRL (mg/kg)	
FP	0226	Apple	2	
FI	0327	Banana	1	
VB	0041	Cabbages, Head	1	
VB	0404	Cauliflower	1	
FS	0013	Cherries	2	
FC	0001	Citrus fruits	2	
JF	0001	Citrus juice	0.2	
VP	0526	Common bean (pods and/ or immature seeds)	0.1	
FB	0269	Grapes	0.5	
VL	0482	Lettuce, Head	2	
MM	0095	Meat	2	(fat) V
ML	0106	Milks	0.05	F V
OC	0305	Olive oil, crude	1	
FT	0305	Olives	1	
VA	0385	Onion, Bulb	0.1	
FS	0247	Peach	2	
FP	0230	Pear	2	
VP	0063	Peas	0.5	
FS	0014	Plums (including Prunes)	1	
VR	0589	Potato	0.05	(*)
GC	0649	Rice	0.1	
VC	0431	Squash, Summer	0.2	
FB	0275	Strawberry	2	
VR	0508	Sweet potato	0.1	
VO	0448	Tomato	0.5	
GC	0654	Wheat	0.1	
VC	0433	Winter squash	0.2	

(40) **FENTIN**

JMPR   65, 70, 72, 86, 91
ADI   0.0005 mg/kg body weight | (1970 and confirmed 1991)
Residue   Fentin, excluding inorganic tin and di- and mono-phenyltin

Commodity code No.	Name	MRL (mg/kg)		
SB 0715	Cacao beans	0.1	(*)	1/
VR 0577	Carrot	0.2		1/
VR 0578	Celeriac	0.1		1/
VS 0624	Celery	1		1/
SB 0716	Coffee beans	0.1	(*)	1/
SO 0697	Peanut	0.05	(*)	1/
TN 0672	Pecan	0.05	(*)	1/
VR 0589	Potato	0.1		
GC 0649	Rice	0.1	(*)	
VR 0596	Sugar beet	0.2		

1/ The Committee decided that if no new information became available
the current CXLs should be recommended for deletion

(41) **FOLPET**

```
JMPR    69, 73, 74, 82, 84, 86, 87, 90
  TADI  0.01 mg/kg body weight  | (TADI 1986-1990 extended to 1993)
Residue  Folpet.
Note     The 24th CCPR decided to maintain the CXLs as temporary for all
         commodities (24.89) irrespective of the status of the ADI.
```

Commodity code No.		Name	TMRL (mg/kg)	
FP	0226	Apple	10	T
FS	0013	Cherries	15	T
FC	0001	Citrus fruits	10	T 1/
VC	0424	Cucumber	2	T
FB	0269	Grapes	25	T
VL	0482	Lettuce, Head	15	T 1/
VC	0046	Melons, except Watermelon	2	T 1/
VA	0385	Onion, Bulb	2	T
FB	0275	Strawberry	20	T
VO	0448	Tomato	5	T

1/ JMPR 1990 proposed to withdraw the CXLs in view of not expected
   uses.

(42) **FORMOTHION**

JMPR   69, 72, 73, 78
ADI   0.02 mg/kg body weight | (1973)
Residue   Formothion
Note   See also Item (27) dimethoate and (55) omethoate.

Commodity code No.	N a m e	MRL (mg/kg)
FC 0001	Citrus fruits	0.2

----

(43) HEPTACHLOR

JMPR   65, 66, 67, 68, 69, 70, 74, 75, 77, 87, 91
ADI   0.0001 mg/kg body weight | (1991)
Residue   Sum of heptachlor and heptachlor epoxide (fat-soluble).

Commodity code No.	Name	MRL (mg/kg)		
VR 0577	Carrot	0.2	E	T 1/
GC 0080	Cereal grains	0.02	E	
FC 0001	Citrus fruits	0.01	E	
SO 0691	Cotton seed	0.02	E	
PE 0112	Eggs	0.05	E	
MM 0095	Meat	0.2	E (fat)	
ML 0106	Milks	0.006	E,F	
FI 0353	Pineapple	0.01	E	
PM 0110	Poultry meat	0.2	E (fat)	
VP 0541	Soya bean (immature seeds)	0.02	E	
OC 0541	Soya bean oil, crude	0.5	E	
OR 0541	Soya bean oil, refined	0.02	E	
VR 0596	Sugar beet	0.05	E	
VO 0448	Tomato	0.02	E	T 1/
A01 0002	Vegetables (except..)	0.05	E	T 1/ 2/

1/ Temporary pending the receipt of further information
2/ (Except as otherwise listed)

(45) **HYDROGEN CYANIDE**

JMPR   65
ADI   0.05 mg/kg body weight | (1965)
Residue   All cyanides, expressed as hydrogen cyanide.

Commodity code No.	N a m e	MRL (mg/kg)	
GC 0080	Cereal grains	75	Po
CF 1211	Wheat flour	6	PoP

----

(46) **HYDROGEN PHOSPHIDE**

JMPR   65, 66, 67, 69, 71
ADI | Not necessary.
Residue   All phosphides, expressed as hydrogen phosphide.
Note   Good usage practices should ensure that residues are not present
          at time of consumption. (JMPR,1966).

Commodity code No.		Name	MRL (mg/kg)	
SB	0715	Cacao beans	0.01	Po
GC	0080	Cereal grains	0.1	Po
DF	0167	Dried fruits	0.01	Po
DV	0168	Dried vegetables	0.01	Po
SO	0697	Peanut	0.01	Po
HS	0093	Spices	0.01	Po
TN	0085	Tree nuts	0.01	Po

(47) INORGANIC BROMIDE

JMPR   68, 69, 71, 79, 81, 83, 88, 89
  ADI   1 mg/kg body weight | (confirmed 1988)
Residue   Bromide ion
Note   MRLs apply to bromide ion from all sources but do not include
           covalently bound bromine (1989 JMPR).

Commodity code No.	Name	MRL (mg/kg)	
FI 0326	Avocado	75	
VB 0041	Cabbages, Head	100	
VS 0624	Celery	300	
GC 0080	Cereal grains	50	
FC 0001	Citrus fruits	30	
VC 0424	Cucumber	50	
DF 0295	Dates, dried or dried & candied	100	
DF 0167	Dried fruits	30	
DF 0269	Dried grapes	100	
DH 0170	Dried herbs	400	
DF 0297	Figs, dried or dried and candied	250	
A02 0002	Fruits (except..)	20	1/
VL 0482	Lettuce, Head	100	
DF 0247	Peach, dried	50	
DF 0014	Prunes	20	
HS 0093	Spices	400	
FB 0275	Strawberry	30	
VO 0448	Tomato	75	
CF 1212	Wheat wholemeal	50	

1/ (Except as otherwise listed)

(48) **LINDANE**

JMPR   65, 66, 67, 68, 69, 70(Annex VI 71 Ev), 71, 73, 74, 75, 77, 78, 79, 89
ADI   0.008 mg/kg body weight | (1989)
Residue   Gamma-HCH (fat-soluble).

	Commodity		MRL (mg/kg)	
code No.	Name			
FP	0226	Apple	0.5	
VD	0071	Beans (dry)	1	Po
VB	0402	Brussels sprouts	0.5	
VB	0403	Cabbage, Savoy	0.5	
VB	0041	Cabbages, Head	0.5	
SB	0715	Cacao beans	1	
VR	0577	Carrot	0.2	E
VB	0404	Cauliflower	0.5	
GC	0080	Cereal grains	0.5	Po
FS	0013	Cherries	0.5	
DM	1215	Cocoa butter	1	
DM	1216	Cocoa mass	1	
FB	0265	Cranberry	3	
FB	0279	Currant, Red, White	0.5	
PE	0112	Eggs	0.1	E
VL	0476	Endive	2	
FB	0269	Grapes	0.5	
VB	0405	Kohlrabi	1	
VL	0482	Lettuce, Head	2	
MM	0097	Meat of cattle, pigs & sheep	2	(fat) V
ML	0106	Milks	0.01	F V
FP	0230	Pear	0.5	
VP	0063	Peas	0.1	
FS	0014	Plums (including Prunes)	0.5	
VR	0589	Potato	0.05	(*)
PM	0110	Poultry meat	0.7	E (fat)
VR	0494	Radish	1	
SO	0495	Rape seed	0.05	(*)
VL	0502	Spinach	2	
FB	0275	Strawberry	3	
VR	0596	Sugar beet	0.1	
AV	0596	Sugar beet leaves or tops	0.1	
VO	0448	Tomato	2	

(49) MALATHION

JMPR   65, 66, 67(corr. to 66 report and eval), 68, 69, 70, 73, 75, 77, 84
ADI   0.02 mg/kg body weight | (1966)
Residue   Malathion.

Commodity code No.	Name	MRL (mg/kg)	
FP 0226	Apple	2	
VD 0071	Beans (dry)	8	Po
FB 0264	Blackberries	8	
FB 0020	Blueberries	0.5	
VB 0400	Broccoli	5	
VB 0041	Cabbages, Head	8	
VB 0404	Cauliflower	0.5	
VS 0624	Celery	1	
GC 0080	Cereal grains	8	Po
VL 0464	Chard	0.5	
FS 0013	Cherries	6	
FC 0001	Citrus fruits	4	
VP 0526	Common bean (pods and/ or immature seeds)	2	
DF 0167	Dried fruits	8	
VO 0440	Egg plant	0.5	
VL 0476	Endive	8	
FB 0269	Grapes	8	
VL 0480	Kale	3	
VB 0405	Kohlrabi	0.5	
VD 0533	Lentil (dry)	8	
VL 0482	Lettuce, Head	8	
AO5 1900	Nuts (whole in shell)	8	
FS 0247	Peach	6	
FP 0230	Pear	0.5	
VP 0063	Peas	0.5	
VO 0051	Peppers	0.5	
FS 0014	Plums (including Prunes)	6	
FB 0272	Raspberries, Red, Black	8	
VR 0075	Root and tuber vegetables	0.5	1/
CM 0650	Rye bran, unprocessed	20	PoP
CF 1250	Rye flour	2	PoP
CF 1251	Rye wholemeal	2	PoP
VL 0502	Spinach	8	
FB 0275	Strawberry	1	
VO 0448	Tomato	3	
VR 0506	Turnip, Garden	3	
CM 0654	Wheat bran, unprocessed	20	PoP

(49) MALATHION

JMPR   65, 66, 67(corr. to 66 report and eval), 68, 69, 70, 73, 75, 77, 84
  ADI   0.02 mg/kg body weight | (1966)
Residue   Malathion.

Commodity code No.	Name	MRL (mg/kg)	
CF 1211	Wheat flour	2	PoP
CF 1212	Wheat wholemeal	2	PoP

1/ (Except turnip, garden)

(51) METHIDATHION

JMPR   72, 75, 79
ADI   0.005 mg/kg body weight | (1975)
Residue   Methidathion (fat-soluble).

Commodity code No.	Name	MRL (mg/kg)	
FP 0226	Apple	0.5	
FS 0240	Apricot	0.2	
VB 0041	Cabbages, Head	0.2	
MF 0812	Cattle fat	0.02	(*)
VB 0404	Cauliflower	0.2	
FS 0013	Cherries	0.2	
VP 0526	Common bean (pods and/ or immature seeds)	0.1	
SO 0691	Cotton seed	0.2	
OC 0691	Cotton seed oil, crude	1	
MO 0097	Edible offal of cattle, pigs and sheep	0.02	(*)
PE 0112	Eggs	0.02	(*)
FB 0269	Grapes	0.2	
DH 1100	Hops, dry	3	
VL 0053	Leafy vegetables	0.2	
FC 0002	Lemons and Limes	2	
GC 0645	Maize	0.1	
FC 0003	Mandarins	5	
MM 0097	Meat of cattle, pigs & sheep	0.02	(*)
ML 0106	Milks	0.0008	(*) F
FS 0245	Nectarine	0.2	
FC 0004	Oranges, Sweet, Sour	2	
FS 0247	Peach	0.2	
FP 0230	Pear	0.5	
VP 0063	Peas	0.1	
MF 0818	Pig fat	0.02	(*)
FS 0014	Plums (including Prunes)	0.2	
VR 0589	Potato	0.02	(*)
PF 0111	Poultry fats	0.02	(*)
PM 0110	Poultry meat	0.02	(*)
PO 0111	Poultry, Edible offal of	0.02	(*)
FC 0005	Shaddocks or Pomelos	2	
MF 0822	Sheep fat	0.02	(*)
GC 0651	Sorghum	0.1	
DT 1114	Tea, Green, Black	0.1	
VO 0448	Tomato	0.1	

(53) **MEVINPHOS**

JMPR   65, 72
  ADI   0.0015 mg/kg body weight | (1972)
Residue   Sum of cis- and trans-mevinphos.
Note   22nd CCPR agreed that additional information was required to support
           CXLs (22.358).

Commodity code No.		N a m e	MRL (mg/kg)
FP	0226	Apple	0.5
FS	0240	Apricot	0.2
VB	0400	Broccoli	1
VB	0402	Brussels sprouts	1
VB	0041	Cabbages, Head	1
VR	0577	Carrot	0.1
VB	0404	Cauliflower	1
FS	0013	Cherries	1
FC	0001	Citrus fruits	0.2
VP	0526	Common bean (pods and/ or immature seeds)	0.1
VC	0424	Cucumber	0.2
FB	0269	Grapes	0.5
VL	0480	Kale	1
VL	0482	Lettuce, Head	0.5
VC	0046	Melons, except Watermelon	0.05
VA	0385	Onion, Bulb	0.1
FS	0247	Peach	0.5
FP	0230	Pear	0.2
VP	0063	Peas	0.1
VR	0589	Potato	0.1
VL	0502	Spinach	0.5
FB	0275	Strawberry	1
VO	0448	Tomato	0.2
VR	0506	Turnip, Garden	0.1

(54) MONOCROTOPHOS

JMPR 72, 75, 91
  ADI 0.00005 mg/kg body weight | (1991)
Residue Monocrotophos.
Note ADI lowered from 0.0006 mg/kg

Commodity code No.	Name	MRL (mg/kg)	
FP 0226	Apple	1	
VB 0402	Brussels sprouts	0.2	
VB 0041	Cabbages, Head	0.2	
VR 0577	Carrot	0.05	(*)
VB 0404	Cauliflower	0.2	
FC 0001	Citrus fruits	0.2	
SB 0716	Coffee beans	0.1	
VP 0526	Common bean (pods and/ or immature seeds)	0.2	
SO 0691	Cotton seed	0.1	
OC 0691	Cotton seed oil, crude	0.05	(*)
MO 0097	Edible offal of cattle, pigs and sheep	0.02	(*)
PE 0112	Eggs	0.02	(*)
MM 0814	Goat meat	0.02	(*)
MO 0814	Goat, Edible offal of	0.02	(*)
DH 1100	Hops, dry	1	
GC 0645	Maize	0.05	(*)
MM 0097	Meat of cattle, pigs & sheep	0.02	(*)
AO3 0001	Milk products	0.02	(*)
ML 0106	Milks	0.002	(*)
VA 0385	Onion, Bulb	0.1	
FP 0230	Pear	1	
VP 0063	Peas	0.1	
VR 0589	Potato	0.05	(*)
PM 0110	Poultry meat	0.02	(*)
PO 0111	Poultry, Edible offal of	0.02	(*)
VP 0541	Soya bean (immature seeds)	0.05	(*)
VR 0596	Sugar beet	0.05	(*)
VO 0448	Tomato	1	
VR 0506	Turnip, Garden	0.05	(*)

(55) OMETHOATE

JMPR	71, 75, 78, 79, 81, 84, 85, 86, 87, 88, 90
ADI	0.0003 mg/kg body weight	(1985)
Residue	Omethoate.
Note	The MRLs apply to residues that may have resulted from the use of formothion, dimethoate or omethoate. (JMPR 1988).
	The Committee agreed that the entries for omethoate should indicate the source of each residue (23.101).

Commodity code No.		Name	MRL (mg/kg)
VS	0620	Artichoke, Globe	0.5
VP	0061	Beans, except broad bean and soya bean	0.2
VB	0400	Broccoli	0.2
VB	0402	Brussels sprouts	0.2
VB	0041	Cabbages, Head	0.2
VR	0577	Carrot	0.05
VB	0404	Cauliflower	0.2
VS	0624	Celery	0.1
GC	0080	Cereal grains	0.05
FC	0001	Citrus fruits	2
VC	0424	Cucumber	0.2
FB	0278	Currant, Black	2
DH	1100	Hops, dry	3
VL	0480	Kale	0.2
VL	0482	Lettuce, Head	0.2
VL	0483	Lettuce, Leaf	0.2
VA	0385	Onion, Bulb	0.1
VP	0063	Peas	0.1
VO	0051	Peppers	1
VR	0589	Potato	0.05
VL	0502	Spinach	0.1
FB	0275	Strawberry	1
VR	0596	Sugar beet	0.05
VO	0448	Tomato	1
VR	0506	Turnip, Garden	0.2

(56) ORTHO-PHENYLPHENOL

JMPR   69, 75, 83, 85, 89, 90
ADI   0.02 mg/kg body weight | (1990)
Residue   Sum of 2-phenylphenol and 2-phenylphenate, expressed as
           2-phenylphenol.
Note   23rd CCPR decided to request information on actual GAP and relevant
           residue data for evaluation by the 1994 JMPR (23.103).

Commodity code No.		Name	MRL (mg/kg)	
FP	0226	Apple	25	Po
VR	0577	Carrot	20	Po
FS	0013	Cherries	3	Po
FC	0001	Citrus fruits	10	Po
VC	0424	Cucumber	10	Po
FS	0245	Nectarine	3	Po
FS	0247	Peach	20	Po
FP	0230	Pear	25	Po
VO	0051	Peppers	10	Po
FI	0353	Pineapple	10	Po
FS	0014	Plums (including Prunes)	15	Po
VR	0508	Sweet potato	15	Po
VO	0448	Tomato	10	Po

(57) PARAQUAT

JMPR   70, 72, 76, 78, 81, 82, 85, 86
ADI   0.004 | mg paraquat cation/kg body weight. (1986)
Residue Paraquat cation (generally available as dichloride, JMPR 1986).
Note   ADI and MRLs are based on data resulting from the use of paraquat
    dichloride (23.105).

Commodity code No.		Name	MRL (mg/kg)	
MO	1280	Cattle, kidney	0.5	
SO	0691	Cotton seed	0.2	
OR	0691	Cotton seed oil, edible	0.05	(*)
MO	0097	Edible offal of cattle, pigs and sheep	0.05	(*) 1/
PE	0112	Eggs	0.01	(*)
DH	1100	Hops, dry	0.2	
GC	0645	Maize	0.1	
MM	0097	Meat of cattle, pigs & sheep	0.05	(*)
ML	0106	Milks	0.01	(*)
FT	0305	Olives	1	
FI	0351	Passion fruit	0.2	
MO	1284	Pig, kidney	0.5	
VR	0589	Potato	0.2	
GC	0649	Rice	10	
CM	1205	Rice, polished	0.5	
MO	1288	Sheep, kidney	0.5	
GC	0651	Sorghum	0.5	
VD	0541	Soya bean (dry)	0.1	
SO	0702	Sunflower seed	2	
OC	0702	Sunflower seed oil, crude	0.05	(*)
OR	0702	Sunflower seed oil, edible	0.05	(*)
AO1	0001	Vegetables	0.05	(*)

1/ (Except as otherwise listed).

(58) PARATHION

JMPR   65, 67, 69, 70, 84
    ADI   0.005 mg/kg body weight | (1967)
Residue   Parathion.

Commodity code No.	N a m e	MRL (mg/kg)	
FS 0240	Apricot	1	
FC 0001	Citrus fruits	1	1/
A02 0002	Fruits (except..)	0.5	1/ 2/
FS 0247	Peach	1	
A01 0002	Vegetables (except..)	0.7	1/

1/ 1991 JMPR recommended to withdraw the MRL

2/ (Except as otherwise listed)

(59) PARATHION-METHYL

JMPR   65, 68, 72, 75, 78, 79, 80, 82, 84
ADI   0.02 mg/kg body weight | (1984)
Residue   Parathion-methyl.

Commodity code No.	Name	MRL (mg/kg)	
VB 0040	Brassica vegetables	0.2	
OC 0691	Cotton seed oil, crude	0.05	
OR 0691	Cotton seed oil, edible	0.05	
VC 0424	Cucumber	0.2	
A02 0001	Fruits	0.2	
DH 1100	Hops, dry	0.05	(*)
VC 0046	Melons, except Watermelon	0.2	
VR 0596	Sugar beet	0.05	(*)
DT 1114	Tea, Green, Black	0.2	
VO 0448	Tomato	0.2	

(60) PHOSALONE

JMPR   72, 75, 76
  ADI   0.006 mg/kg body weight | (1972)
Residue   Phosalone.

Commodity code No.	Name	MRL (mg/kg)
FP 0226	Apple	5
VR 0574	Beetroot	2
VB 0400	Broccoli	1
VB 0402	Brussels sprouts	1
VB 0041	Cabbages, Head	1
FS 0013	Cherries	10
TN 0664	Chestnuts	0.1 (*)
FC 0001	Citrus fruits	1
VC 0424	Cucumber	1
FB 0269	Grapes	5
DH 1100	Hops, dry	2
VL 0482	Lettuce, Head	1
FS 0247	Peach	5
FP 0230	Pear	2
VP 0063	Peas	1
TN 0672	Pecan	0.1 (*)
FS 0014	Plums (including Prunes)	5
VR 0589	Potato	0.1 (*)
SO 0495	Rape seed	0.1 (*)
MF 0822	Sheep fat	0.5 V
MM 0822	Sheep meat	0.05 (*) V
FB 0275	Strawberry	1
VO 0448	Tomato	1

(61) PHOSPHAMIDON

JMPR   65, 66, 68, 69, 72, 74, 82, 85, 86
ADI   0.0005 mg/kg body weight | (1986)
Residue Sum of phosphamidon (E- and Z- isomers) and N-desethyl-phosphamidon
         (E- and Z- isomers).

Commodity code No.		Name	MRL (mg/kg)	
FP	0226	Apple	0.5	
VB	0400	Broccoli	0.2	
VB	0402	Brussels sprouts	0.2	
VB	0041	Cabbages, Head	0.2	
VR	0577	Carrot	0.2	
VR	0578	Celeriac	0.2	
GC	0080	Cereal grains	0.1	
FS	0013	Cherries	0.2	
FC	0001	Citrus fruits	0.4	
VP	0526	Common bean (pods and/ or immature seeds)	0.2	
VC	0424	Cucumber	0.1	
VL	0482	Lettuce, Head	0.1	
FS	0247	Peach	0.2	
FP	0230	Pear	0.5	
VP	0063	Peas	0.2	
VO	0051	Peppers	0.2	
FS	0014	Plums (including Prunes)	0.2	
VR	0075	Root and tuber vegetables	0.05	(*) 1/
VL	0502	Spinach	0.2	
FB	0275	Strawberry	0.2	
VO	0448	Tomato	0.1	
VC	0432	Watermelon	0.1	

1/ (Except carrot and celeriac).

(62) **PIPERONYL BUTOXIDE**

JMPR   65, 66, 67, 69, 72
ADI   0.03 mg/kg body weight | (1972)
Residue   Piperonyl butoxide.

Commodity code No.	Name	MRL (mg/kg)	
GC 0080	Cereal grains	20	Po
MD 0180	Dried fish	20	Po
DF 0167	Dried fruits	8	Po
DV 0168	Dried vegetables	8	Po
SO 0089	Oilseed, except peanut	8	Po
SO 0697	Peanut	8	Po T
TN 0085	Tree nuts	8	Po

(63) PYRETHRINS

JMPR   65, 66, 67, 68, 69, 70, 72, 74
ADI   0.04 mg/kg body weight | (1972)
Residue   Sum of pyrethrins I and II, cinerins I and II, and jasmolins I and II,
          determined after calibration by means of the International
          Pyrethrum Standard.

Commodity code No.		Name	MRL (mg/kg)	
GC	0080	Cereal grains	3	Po
MD	0180	Dried fish	3	Po
DF	0167	Dried fruits	1	Po
DV	0168	Dried vegetables	1	Po
SO	0088	Oilseed	1	Po
TN	0085	Tree nuts	1	Po

(64) **QUINTOZENE**

JMPR   69, 73, 74, 75, 76(annex I, correction to 75 report), 77
ADI   0.007 mg/kg body weight | (confirmed 1977)
Residue   Sum of quintozene, pentachloroaniline and methyl pentachlorophenyl
       sulphide.

Commodity code No.	Name	MRL (mg/kg)
FI 0327	Banana	1
VB 0400	Broccoli	0.02
VB 0041	Cabbages, Head	0.02
VP 0526	Common bean (pods and/ or immature seeds)	0.01
VD 0526	Common bean, dry	0.2
SO 0691	Cotton seed	0.03
VL 0482	Lettuce, Head	3
SO 0697	Peanut	2
SO 0703	Peanut, whole	5
VO 0445	Peppers, Sweet	0.01
VR 0589	Potato	0.2
VO 0448	Tomato	0.1

(65) THIABENDAZOLE

JMPR   70, 71, 72, 75, 77, 79, 81
ADI   0.3 mg/kg body weight | (1977)
Residue   Thiabendazole, or, in the case of animal products, sum of thiabendazole and 5-hydroxythiabendazole.

Commodity code No.		Name	MRL (mg/kg)	
FP	0226	Apple	10	
FI	0327	Banana	3	
GC	0080	Cereal grains	0.2	
FC	0001	Citrus fruits	10	Po
MO	0096	Edible offal of cattle, goats,horses,pigs & sheep	0.1	(*)
MM	0096	Meat of cattle,goats, horses,pigs /sheep	0.1	(*)
ML	0106	Milks	0.1	(*)
VA	0385	Onion, Bulb	0.1	
FP	0230	Pear	10	
VR	0589	Potato	5	Po 1/
FB	0275	Strawberry	3	
VR	0596	Sugar beet	5	
AV	0596	Sugar beet leaves or tops	10	
DM	0596	Sugar beet molasses	1	
AB	0596	Sugar beet pulp, dry	5	
VO	0448	Tomato	2	

1/ Washed before analysis

(66) **TRICHLORFON**

```
JMPR    71, 75, 78, 87
  ADI   0.01 mg/kg body weight  | (1978)
Residue  Trichlorfon.
Note     Scheduled for periodic review (24.241).
```

	Commodity		MRL (mg/kg)	
	code No.	N a m e		
FP	0226	Apple	2	
VS	0620	Artichoke, Globe	0.1	
FI	0327	Banana	1	
VR	0574	Beetroot	0.2	
VB	0402	Brussels sprouts	0.2	
VB	0041	Cabbages, Head	0.5	
VR	0577	Carrot	0.05	
MF	0812	Cattle fat	0.1	
MO	0812	Cattle, Edible offal of	0.1	
VB	0404	Cauliflower	0.2	
VS	0624	Celery	0.2	
GC	0080	Cereal grains	0.1	
FS	0013	Cherries	0.1	
FC	0001	Citrus fruits	0.1	
VP	0526	Common bean (pods and/ or immature seeds)	0.1	
SO	0691	Cotton seed	0.1	
VD	0527	Cowpea (dry)	0.1	
VO	0440	Egg plant	0.05	
FB	0269	Grapes	0.5	
VL	0480	Kale	0.2	
VL	0482	Lettuce, Head	0.5	
VP	0534	Lima bean (young pods / immature beans)	0.1	
SO	0693	Linseed	0.1	
MM	0097	Meat of cattle, pigs & sheep	0.1	V
ML	0106	Milks	0.05	V
VL	0485	Mustard greens	0.1	
HH	0740	Parsley	0.05	
FS	0247	Peach	0.2	
SO	0697	Peanut	0.1	
VO	0051	Peppers	1	
MF	0818	Pig fat	0.1	
MO	0818	Pig, Edible offal of	0.1	
VC	0429	Pumpkins	0.1	
VR	0494	Radish	0.1	
SO	0495	Rape seed	0.1	
SO	0699	Safflower seed	0.1	

(66) **TRICHLORFON**

JMPR   71, 75, 78, 87
  ADI   0.01 mg/kg body weight | (1978)
Residue   Trichlorfon.
Note   Scheduled for periodic review (24.241).

Commodity code No.	Name	MRL (mg/kg)
VD 0541	Soya bean (dry)	0.1
VL 0502	Spinach	0.5
FB 0275	Strawberry	1
VR 0596	Sugar beet	0.05
VO 0447	Sweet corn (corn-on-the-cob)	0.2
VO 0448	Tomato	0.2
VR 0506	Turnip, Garden	0.1

(67) **CYHEXATIN**

JMPR   70, 73, 74, 75, 77, 78, 80, 81, 82, 83, 85, 88, 89, 91
  ADI   0.001 mg/kg body weight | (1991)
Residue   Cyhexatin.

Commodity code No.	Name	MRL (mg/kg)		
FP 0226	Apple	2		
FC 0001	Citrus fruits	2		
VC 0424	Cucumber	0.5		
VC 0425	Gherkin	1		
MM 0095	Meat	0.2	V	
VC 0046	Melons, except Watermelon	0.5		
A03 0001	Milk products	0.05	(*) V	
ML 0106	Milks	0.05	(*) V	
FP 0230	Pear	2		
VO 0445	Peppers, Sweet	0.5		
DT 1114	Tea, Green, Black	2		6/
VO 0448	Tomato	2		

6/ Proposed deletion by the 1991 JMPR

(70) BROMOPROPYLATE

JMPR   73
  ADI   0.008 mg/kg body weight | (1973)
Residue   Bromopropylate.

Commodity code No.		Name	MRL (mg/kg)
FP	0226	Apple	5
FI	0327	Banana	5
FS	0013	Cherries	5
FC	0001	Citrus fruits	5
SO	0691	Cotton seed	1
FB	0269	Grapes	5
DH	1100	Hops, dry	5
FS	0245	Nectarine	5
FS	0247	Peach	5
FP	0230	Pear	5
FS	0014	Plums (including Prunes)	5
FB	0275	Strawberry	5
DT	1114	Tea, Green, Black	5
A01	0001	Vegetables	1

----

(72) CARBENDAZIM

JMPR   73, 76, 77, 78, 83, 85, 87, 88, 90
ADI   0.01 mg/kg body weight | (1983)
Residue   Carbendazim.
Note   MRLs cover carbendazim residues occurring as a metabolic product of
benomyl or thiophanate-methyl, or from direct use of carbendazim.
Data bases: B = benomyl; C = carbendazim; Th = thiophanate-methyl.

Commodity code No.		Name	MRL (mg/kg)		
VS	0621	Asparagus	0.1	(*)	B
FI	0326	Avocado	0.5		B
FI	0327	Banana	1	Po	B,C,Th
AS	0640	Barley straw and fodder, dry	2		B
VD	0071	Beans (dry)	2		B
VP	0522	Broad bean (green pods/ immature seeds)	2		Th
VB	0402	Brussels sprouts	0.5		B
MM	0812	Cattle meat	0.1	(*)	B
VS	0624	Celery	2		B,C
PF	0840	Chicken fat	0.1	(*)	Th
SB	0716	Coffee beans	0.1	(*)	C
VP	0526	Common bean (pods and/ or immature seeds)	2		B,C,Th
VC	0424	Cucumber	0.5		B,C,Th
VO	0440	Egg plant	0.5		C
PE	0112	Eggs	0.1	(*)	B, Th
VC	0425	Gherkin	2		C,Th
DH	1100	Hops, dry	50		C
FI	0345	Mango	2		B
VC	0046	Melons, except Watermelon	2	Po,	B,C
ML	0106	Milks	0.1	(*)	B
VA	0385	Onion, Bulb	2		C,Th
SO	0697	Peanut	0.1	(*)	B,C
AL	0697	Peanut fodder	5		B,C
VR	0589	Potato	3	Po,	B,C 5/
PM	0110	Poultry meat	0.1	(*)	B,Th
SO	0495	Rape seed	0.1	(*)	C
AS	0649	Rice straw and fodder, dry	15		B,C,Th
MM	0822	Sheep meat	0.1	(*)	B
VD	0541	Soya bean (dry)	0.2		C
AL	0541	Soya bean fodder	0.1	(*)	C
VC	0431	Squash, Summer	0.5		B
VR	0596	Sugar beet	0.1	(*)	B,C,Th
VR	0497	Swede	0.1	(*)	C
VR	0508	Sweet potato	1		B

(72) **CARBENDAZIM**

JMPR   73, 76, 77, 78, 83, 85, 87, 88, 90
ADI   0.01 mg/kg body weight | (1983)
Residue   Carbendazim.
Note   MRLs cover carbendazim residues occurring as a metabolic product of
benomyl or thiophanate-methyl, or from direct use of carbendazim.
Data bases: B = benomyl; C = carbendazim; Th = thiophanate-methyl.

Commodity code No.	N a m e	MRL (mg/kg)		
VR 0505	Taro	0.1	(*)	B
TN 0085	Tree nuts	0.1	(*)	B
AS 0654	Wheat straw and fodder, dry	5		B
VC 0433	Winter squash	0.5		B

5/ (washed before analysis)

(74) DISULFOTON

JMPR   73, 75, 79, 81, 84, 91
ADI   0.0003 mg/kg body weight | (1991)
Residue   Sum of disulfoton, demeton-S and their sulphoxides and sulphones,
         expressed as disulfoton.

Commodity code No.	Name	MRL (mg/kg)	
AL 1020	Alfalfa fodder	10	
VS 0624	Celery	0.5	
GC 0081	Cereal grain (except....)	0.2	1/ 2/
AL 1031	Clover hay or fodder	10	
SB 0716	Coffee beans	0.1	3/
A03 1600	Forage creps (green)	5	2/
GC 0645	Maize	0.5	
SO 0697	Peanut	0.1	3/
TN 0672	Pecan	0.1	3/
FI 0353	Pineapple	0.1	3/
VR 0589	Potato	0.5	
GC 0649	Rice	0.5	
VD 0541	Soya bean (dry)	0.1	3/
VR 0596	Sugar beet	0.5	
A01 0001	Vegetables	0.5	2/

1/ (Except rice and maize)

2/ JMPR 1991 recommended to withdraw the MRL

3/ 1991 JMPR established the limit of determination at 0.01 mg/kg

(75) PROPOXUR

JMPR   73, 77, 81, 83, 89, 91
ADI   0.02 mg/kg body weight | (1973 and confirmed 1989)
Residue   Propoxur.

Commodity code No.	Name	MRL (mg/kg)		
FP 0226	Apple	3		
FB 0264	Blackberries	3		
FS 0013	Cherries	3		
FB 0279	Currant, Red, White	3		
FB 0268	Gooseberry	3		
AL 0157	Legume animal feeds	1	fresh wt	
MM 0095	Meat	0.05	(*)	
ML 0106	Milks	0.05	(*)	
FS 0247	Peach	3		
FP 0230	Pear	3		
FS 0014	Plums (including Prunes)	3		
CM 0649	Rice, husked	0.1		
VR 0075	Root and tuber vegetables	0.5		1/
FB 0275	Strawberry	3		
A01 0001	Vegetables	3		1/

1/ 1991 JMPR recommended to withdraw the MRL

(76) THIOMETON

JMPR   69, 73, 76, 79, 88
ADI   0.003 mg/kg body weight | (1979)
Residue Sum of thiometon, its sulphoxide and sulphone, expressed as thiometon.
Note   Scheduled for periodic review (24.241).

Commodity code No.	Name	MRL (mg/kg)	
FP 0226	Apple	0.5	
FS 0240	Apricot	0.5	
VB 0041	Cabbages, Head	0.5	
VR 0577	Carrot	0.05	(*)
VS 0624	Celery	0.5	
GC 0080	Cereal grains	0.05	(*)
FS 0244	Cherry, Sweet	0.5	
VL 0469	Chicory leaves	0.5	
VP 0526	Common bean (pods and/ or immature seeds)	0.5	
OC 0691	Cotton seed oil, crude	0.1	(*)
VO 0440	Egg plant	0.5	
VL 0476	Endive	0.5	
AM 1051	Fodder beet	0.05	(*)
AV 1051	Fodder beet leaves or tops	0.05	(*)
FB 0269	Grapes	0.5	
DH 1100	Hops, dry	2	
VL 0482	Lettuce, Head	0.5	
AF 0645	Maize forage	0.1	(*) fresh wt
SO 0090	Mustard seed	0.05	(*)
HH 0740	Parsley	0.5	
FS 0247	Peach	0.5	
SO 0703	Peanut, whole	0.5	
FP 0230	Pear	0.5	
VP 0063	Peas	0.5	
VO 0051	Peppers	0.5	
FS 0014	Plums (including Prunes)	0.5	
VR 0589	Potato	0.05	(*)
FP 0231	Quince	0.5	
SO 0495	Rape seed	0.05	(*)
AS 0081	Straw and fodder (dry) of cereal grains	0.1	(*)
FB 0275	Strawberry	0.5	
VR 0596	Sugar beet	0.05	(*)
AV 0596	Sugar beet leaves or tops	0.05	(*)
VO 0448	Tomato	0.5	

(77) THIOPHANATE-METHYL

JMPR   73, 75, 77, 78, 88, 90
 ADI   0.08 mg/kg body weight | (Confirmed 1977)
Residue   Thiophanate-methyl expressed as carbendazim.
Note   CCPR will recomend to the Commission the deletion of all CXLs
          when the MRLs for carbendazim (72) reach Step 8 (21.112: 22.119).

Commodity code No.	Name	MRL (mg/kg)	
FP 0226	Apple	5	Po
VR 0577	Carrot	5	Po
VS 0624	Celery	20	Po 1/
GC 0080	Cereal grains	0.1	(*)
FS 0013	Cherries	10	
PM 0840	Chicken meat	0.1	(*)
FC 0001	Citrus fruits	10	Po
FB 0278	Currant, Black	5	
FB 0268	Gooseberry	5	
FB 0269	Grapes	10	
VL 0482	Lettuce, Head	5	
VO 0450	Mushrooms	1	
FS 0247	Peach	10	Po
FP 0230	Pear	5	Po
FS 0014	Plums (including Prunes)	2	
FB 0272	Raspberries, Red, Black	5	
FB 0275	Strawberry	5	
AV 0596	Sugar beet leaves or tops	5	
VO 0448	Tomato	5	

1/ The Committee recommended deletion of this CXL unless data to
   support this higher limit respect to the corresponding MRL for
   carbendazim are provided (23.114).

(78) VAMIDOTHION

JMPR   73, 82, 85, 87, 88, 90
   ADI   0.008 mg/kg body weight | (1988)
Residue   Sum of vamidothion, its sulphoxide and sulphone, expressed as
     vamidothion.

Commodity code No.		Name	MRL (mg/kg)
GC	0080	Cereal grains	0.2
FB	0269	Grapes	0.5
FS	0247	Peach	0.5
CM	0649	Rice, husked	0.2
VR	0596	Sugar beet	0.5

(80) CHINOMETHIONAT

JMPR   68, 74, 77, 81, 83, 84, 87
ADI   0.006 mg/kg body weight | (1987)
Residue   Chinomethionat.

Commodity code No.	Name	MRL (mg/kg)	
TN 0660	Almonds	0.1	
FP 0226	Apple	0.2	
FI 0326	Avocado	0.1	
GC 0080	Cereal grains	0.1	
FC 0001	Citrus fruits	0.5	
VC 0424	Cucumber	0.1	
FB 0021	Currants, Black, Red, White	0.1	
VC 0425	Gherkin	0.1	
FB 0268	Gooseberry	0.1	
FB 0269	Grapes	0.1	
TN 0669	Macadamia nuts	0.02	(*)
MM 0095	Meat	0.05	(*)
VC 0046	Melons, except Watermelon	0.1	
ML 0106	Milks	0.01	(*)
FI 0350	Papaya	5	
FT 0307	Persimmon, Japanese	0.05	
FB 0275	Strawberry	0.2	
VC 0432	Watermelon	0.02	

(81) CHLOROTHALONIL

JMPR   74, 77, 78, 79, 81, 83, 84(corr. to 83 report), 85, 87, 88, 90
ADI   0.03 mg/kg body weight | (1990)
Residue   Chlorothalonil.

Commodity code No.	N a m e	MRL (mg/kg)
FI 0327	Banana	0.2
FB 0264	Blackberries	10
VB 0400	Broccoli	5
VB 0402	Brussels sprouts	5
VB 0041	Cabbages, Head	5
VR 0577	Carrot	1
VB 0404	Cauliflower	5
VS 0624	Celery	15
GC 0080	Cereal grains	0.2
FS 0013	Cherries	10
FC 0001	Citrus fruits	5
VP 0526	Common bean (pods and/ or immature seeds)	5
FB 0265	Cranberry	5
VC 0424	Cucumber	5
FB 0021	Currants, Black, Red, White	25
VL 0476	Endive	10
VL 0480	Kale	10
VL 0482	Lettuce, Head	10
VD 0534	Lima bean (dry)	0.5
VC 0046	Melons, except Watermelon	5
VA 0385	Onion, Bulb	5
FS 0247	Peach	25
SO 0697	Peanut	0.1
SO 0703	Peanut, whole	0.5
VO 0051	Peppers	10
VR 0589	Potato	0.1
VC 0429	Pumpkins	5
FB 0272	Raspberries, Red, Black	10
VC 0431	Squash, Summer	5
VR 0596	Sugar beet	1
VO 0447	Sweet corn (corn-on-the-cob)	1
VO 0448	Tomato	5
VC 0433	Winter squash	5
VS 0469	Witloof chicory (sprouts)	10

(82) **DICHLOFLUANID**

JMPR   69, 74, 77, 79, 81, 82, 83, 85
  ADI   0.3 mg/kg body weight | (1983)
Residue   Dichlofluanid.

	Commodity		MRL (mg/kg)
code No.		N a m e	
FP	0226	Apple	5
GC	0640	Barley	0.1
FB	0264	Blackberries	10
FS	0013	Cherries	2
VP	0526	Common bean (pods and/ or immature seeds)	2
VC	0424	Cucumber	5
FB	0021	Currants, Black, Red, White	15
VO	0440	Egg plant	1
FB	0268	Gooseberry	7
FB	0269	Grapes	15
VL	0482	Lettuce, Head	10
GC	0647	Oats	0.1
VA	0385	Onion, Bulb	0.1
FS	0247	Peach	5
FP	0230	Pear	5
VO	0051	Peppers	2
VR	0589	Potato	0.1
FB	0272	Raspberries, Red, Black	15
GC	0650	Rye	0.1
FB	0275	Strawberry	10
VO	0448	Tomato	2
GC	0654	Wheat	0.1
AS	0654	Wheat straw and fodder, dry	0.5

(83) **DICLORAN**

JMPR   74, 77
ADI   0.03 mg/kg body weight | (1977)
Residue   Dichloran.

Commodity code No.	Name	MRL (mg/kg)	
FS 0240	Apricot	10	Po
FB 0264	Blackberries	5	
VR 0577	Carrot	10	Po
FS 0013	Cherries	15	Po
VP 0526	Common bean (pods and/ or immature seeds)	2	
FB 0021	Currants, Black, Red, White	5	
VC 0425	Gherkin	0.5	
FB 0269	Grapes	10	Po
VL 0482	Lettuce, Head	10	
FS 0245	Nectarine	10	Po
VA 0385	Onion, Bulb	10	Po
FS 0247	Peach	15	Po
FS 0014	Plums (including Prunes)	10	Po
FB 0272	Raspberries, Red, Black	10	
FB 0275	Strawberry	10	
VO 0448	Tomato	0.5	
VS 0469	Witloof chicory (sprouts)	1	

(84) **DODINE**

JMPR   74,   76, 77
  ADI   0.01 mg/kg body weight | (1976)
Residue Dodine.

Commodity code No.	Name	MRL (mg/kg)
FP 0226	Apple	5
FS 0013	Cherries	2
FB 0269	Grapes	5
FS 0247	Peach	5
FP 0230	Pear	5
FB 0275	Strawberry	5

(85) FENAMIPHOS

JMPR   74, 77, 78, 80, 85, 87
ADI   0.0005 mg/kg body weight | (1987)
Residue Sum of fenamiphos, its sulphoxide and sulphone, expressed as
         fenamiphos.

Commodity code No.	Name	MRL (mg/kg)	
FI 0327	Banana	0.1	
VB 0400	Broccoli	0.05	(*)
VB 0402	Brussels sprouts	0.05	(*)
VB 0041	Cabbages, Head	0.05	(*)
VR 0577	Carrot	0.2	
VB 0404	Cauliflower	0.05	(*)
SB 0716	Coffee beans	0.1	
SM 0716	Coffee beans, roasted	0.1	
SO 0691	Cotton seed	0.05	(*)
FB 0269	Grapes	0.1	
FI 0341	Kiwifruit	0.05	(*)
VC 0046	Melons, except Watermelon	0.05	(*)
FC 0004	Oranges, Sweet, Sour	0.5	
SO 0697	Peanut	0.05	(*)
FI 0353	Pineapple	0.05	(*)
VR 0589	Potato	0.2	
VD 0541	Soya bean (dry)	0.05	(*)
VR 0596	Sugar beet	0.05	(*)
VR 0508	Sweet potato	0.1	
VO 0448	Tomato	0.2	

(86) PIRIMIPHOS-METHYL

```
JMPR    74, 76, 77, 79, 83, 85
  ADI   0.01 mg/kg body weight  | (1976)
Residue   Pirimiphos-methyl (fat soluble).
```

Commodity code No.	Name	MRL (mg/kg)	
FP 0226	Apple	2	
VB 0402	Brussels sprouts	2	
VB 0041	Cabbages, Head	2	
VR 0577	Carrot	1	
VB 0404	Cauliflower	2	
GC 0080	Cereal grains	10	Po
FS 0013	Cherries	2	
FC 0001	Citrus fruits	2	
VP 0526	Common bean (pods and/ or immature seeds)	0.5	
VC 0424	Cucumber	1	
FB 0278	Currant, Black	1	
DF 0295	Dates, dried or dried & candied	0.5	Po
MD 0180	Dried fish	8	Po
PE 0112	Eggs	0.05	(*)
FB 0268	Gooseberry	1	
FI 0341	Kiwifruit	2	
VL 0482	Lettuce, Head	5	
MM 0095	Meat	0.05	(*)
ML 0106	Milks	0.05	(*) F
VO 0450	Mushrooms	5	
FT 0305	Olives	5	
SO 0697	Peanut	2	Po
OC 0697	Peanut oil, crude	15	PoP
SO 0703	Peanut, whole	25	Po
FP 0230	Pear	2	
VP 0063	Peas	0.05	(*)
VO 0051	Peppers	1	
FS 0014	Plums (including Prunes)	2	
VR 0589	Potato	0.05	(*)
FB 0272	Raspberries, Red, Black	1	
CM 1206	Rice bran, unprocessed	20	PoP
CM 0649	Rice, husked	2	PoP
CM 1205	Rice, polished	1	PoP
CF 1251	Rye wholemeal	5	PoP
VL 0502	Spinach	5	
VA 0389	Spring onion	1	
FB 0275	Strawberry	1	

(86) **PIRIMIPHOS-METHYL**

JMPR   74, 76, 77, 79, 83, 85
ADI   0.01 mg/kg body weight | (1976)
Residue   Pirimiphos-methyl (fat soluble).

Commodity code No.	Name	MRL (mg/kg)	
VO 0448	Tomato	1	
CM 0654	Wheat bran, unprocessed	20	PoP
CF 1211	Wheat flour	2	PoP
CF 1212	Wheat wholemeal	5	PoP
CP 1211	White bread	0.5	PoP
CP 1212	Wholemeal bread	1	PoP

(90) CHLORPYRIFOS-METHYL

JMPR   75, 76(amendment to 75 report in annex I), 79, 90, 91
ADI   0.001 mg/kg body weight | (1991)
Residue   Chlorpyrifos-methyl (fat soluble).

Commodity code No.		Name	MRL (mg/kg)	
FP	0226	Apple	0.5	
VS	0620	Artichoke, Globe	0.1	
VB	0041	Cabbages, Head	0.1	
MF	0812	Cattle fat	0.05	
MM	0812	Cattle meat	0.05	
MO	0812	Cattle, Edible offal of	0.05	
PF	0840	Chicken fat	0.05	
PM	0840	Chicken meat	0.05	
PO	0840	Chicken, Edible offal of	0.05	
VL	0467	Chinese cabbage, type "Pe-tsai"	0.1	
VP	0526	Common bean (pods and/ or immature seeds)	0.1	
VO	0440	Egg plant	0.1	
PE	0112	Eggs	0.05	
VL	0482	Lettuce, Head	0.1	
GC	0645	Maize	10	Po
ML	0106	Milks	0.01	(*) F
FS	0247	Peach	0.5	
VO	0051	Peppers	0.1	
VR	0494	Radish	0.1	
GC	0649	Rice	0.1	
GC	0651	Sorghum	10	Po
DT	1114	Tea, Green, Black	0.1	
VO	0448	Tomato	0.5	
GC	0654	Wheat	10	Po
CM	0654	Wheat bran, unprocessed	20	PoP
CF	1211	Wheat flour	2	PoP
CP	1211	White bread	0.5	PoP
CP	1212	Wholemeal bread	2	PoP

(94) **METHOMYL**

JMPR   75, 76, 77, 78, 86, 87, 88, 89, 90
ADI   0.03 mg/kg body weight | (1989)
Residue   Sum of methomyl and methyl hydroxythioacetimidate ("methomyl oxime"),
expressed as methomyl (JMPR 88).

Note   22nd CCPR decided to combine MRLs for thiodicarb and methomyl
into a single list (22.126). In case of different MRLs the higher
limit would prevail. The list should contain an indication as to
which compound the MRL was based (23.126).

Commodity code No.		Name	MRL (mg/kg)	
AL	1021	Alfalfa forage (green)	10	fresh wt
VS	0621	Asparagus	2	
GC	0640	Barley	0.5	
AS	0640	Barley straw and fodder, dry	5	
VD	0071	Beans (dry)	0.1	
VB	0041	Cabbages, Head	5	
VB	0404	Cauliflower	2	
VS	0624	Celery	2	
FC	0001	Citrus fruits	1	
VP	0526	Common bean (pods and/ or immature seeds)	2	
SO	0691	Cotton seed	0.1	
VC	0424	Cucumber	0.2	
VO	0440	Egg plant	0.2	
FB	0269	Grapes	5	
DH	1100	Hops, dry	10	
VL	0480	Kale	5	
VL	0482	Lettuce, Head	5	
MM	0095	Meat	0.02	(*)
VC	0046	Melons, except Watermelon	0.2	
ML	0106	Milks	0.02	(*)
AM	0738	Mint hay	2	
FS	0245	Nectarine	5	
AS	0647	Oat straw and fodder, dry	5	
GC	0647	Oats	0.5	
VA	0385	Onion, Bulb	0.2	
VA	0387	Onion, Welsh	0.5	
AL	0528	Pea vines (green)	10	fresh wt
FS	0247	Peach	5	
SO	0697	Peanut	0.1	
AL	1270	Peanut forage (green)	5	
VP	0063	Peas	5	
VP	0064	Peas, shelled	0.5	

(94) METHOMYL

JMPR   75, 76, 77, 78, 86, 87, 88, 89, 90
ADI   0.03 mg/kg body weight | (1989)

Residue   Sum of methomyl and methyl hydroxythioacetimidate ("methomyl oxime"),
           expressed as methomyl (JMPR 88).

Note   22nd CCPR decided to combine MRLs for thiodicarb and methomyl
           into a single list (22.126). In case of different MRLs the higher
           limit would prevail. The list should contain an indication as to
           which compound the MRL was based (23.126).

Commodity code No.		Name	MRL (mg/kg)
VO	0051	Peppers	1
FI	0353	Pineapple	0.2
FP	0009	Pome fruits	2
VR	0589	Potato	0.1
GC	0651	Sorghum	0.2
AF	0651	Sorghum forage (green)	1
VP	0541	Soya bean (immature seeds)	0.1
AL	1265	Soya bean forage (green)	10
VL	0502	Spinach	5
VC	0431	Squash, Summer	0.2
VR	0596	Sugar beet	0.1
VO	0447	Sweet corn (corn-on-the-cob)	0.1
VO	0448	Tomato	0.5
VC	0432	Watermelon	0.2
GC	0654	Wheat	0.5
AS	0654	Wheat straw and fodder, dry	5

(95) ACEPHATE

```
JMPR    76, 79, 81, 82, 84, 87, 88, 90
 ADI    0.03 mg/kg body weight  | (1988 and confirmed 1990)
```

Residue   Acephate (the metabolite O,S-dimethyl phosphoramidothioate is
methamidophos (100), which has separate recommendations).

Commodity code No.		Name	MRL (mg/kg)	
AL	1021	Alfalfa forage (green)	10	fresh wt
MF	0812	Cattle fat	0.1	
MM	0812	Cattle meat	0.1	
SO	0691	Cotton seed	2	
PE	0112	Eggs	0.1	
VL	0482	Lettuce, Head	5	
ML	0106	Milks	0.1	
MF	0818	Pig fat	0.1	
MM	0818	Pig meat	0.1	
VR	0589	Potato	0.5	
PF	0111	Poultry fats	0.1	
PM	0110	Poultry meat	0.1	
VD	0541	Soya bean (dry)	0.5	
VR	0596	Sugar beet	0.1	
AV	0596	Sugar beet leaves or tops	10	
FT	0312	Tree tomato	0.5	

(96) **CARBOFURAN**

JMPR   76, 79, 80, 82, 91
ADI   0.01 mg/kg body weight | (Confirmed 1982)
Residue   Sum of carbofuran and 3-hydroxy-carbofuran expressed as carbofuran.

	Commodity		MRL (mg/kg)	
	code No.	N a m e		
AL	1020	Alfalfa fodder	20	
AL	1021	Alfalfa forage (green)	5	
FI	0327	Banana	0.1	(*)
GC	0640	Barley	0.1	(*)
VB	0402	Brussels sprouts	2	
VB	0041	Cabbages, Head	0.5	
VR	0577	Carrot	0.5	
MF	0812	Cattle fat	0.05	(*)
VB	0404	Cauliflower	0.2	
SB	0716	Coffee beans	0.1	(*)
MO	0096	Edible offal of cattle, goats,horses,pigs & sheep	0.05	(*)
VO	0440	Egg plant	0.1	(*)
MF	0814	Goat fat	0.05	(*)
DH	1100	Hops, dry	5	
MF	0816	Horse fat	0.05	(*)
VB	0405	Kohlrabi	0.1	(*)
VL	0482	Lettuce, Head	0.1	(*)
GC	0645	Maize	0.1	(*)
AS	0645	Maize fodder	5	fresh wt
MM	0096	Meat of cattle,goats, horses,pigs /sheep	0.05	(*)
ML	0106	Milks	0.05	(*)
SO	0090	Mustard seed	0.1	(*)
GC	0647	Oats	0.1	(*)
SO	0088	Oilseed	0.1	(*)
VA	0385	Onion, Bulb	0.1	(*)
FS	0247	Peach	0.1	(*)
FP	0230	Pear	0.1	(*)
MF	0818	Pig fat	0.05	(*)
VR	0589	Potato	0.5	
CM	0649	Rice, husked	0.2	
MF	0822	Sheep fat	0.05	(*)
GC	0651	Sorghum	0.1	(*)
VD	0541	Soya bean (dry)	0.2	
FB	0275	Strawberry	0.1	(*)
VR	0596	Sugar beet	0.1	(*)
AV	0596	Sugar beet leaves or tops	0.2	
GS	0659	Sugar cane	0.1	(*)

(96) **CARBOFURAN**

JMPR   76, 79, 80, 82, 91
ADI   0.01 mg/kg body weight | (Confirmed 1982)
Residue   Sum of carbofuran and 3-hydroxy-carbofuran expressed as carbofuran.

Commodity code No.	Name	MRL (mg/kg)	
VO 1275	Sweet corn (kernels)	0.1	(*)
VO 0448	Tomato	0.1	(*)
GC 0654	Wheat	0.1	(*)

(97) **CARTAP**

JMPR   76, 78
ADI   0.1 mg/kg body weight | (1978)
Residue   Cartap, expressed as free base.

	Commodity code No.	N a m e	MRL (mg/kg)
VB	0041	Cabbages, Head	0.2
TN	0664	Chestnuts	0.1
VL	0467	Chinese cabbage, type "Pe-tsai"	2
HS	0784	Ginger, root	0.1
FB	0269	Grapes	1
DH	1100	Hops, dry	5
FT	0307	Persimmon, Japanese	1
VR	0589	Potato	0.1
VR	0494	Radish	1
CM	0649	Rice, husked	0.1
VO	0447	Sweet corn (corn-on-the-cob)	0.1
DT	1114	Tea, Green, Black	20

(99) EDIFENPHOS

JMPR   76, 79, 81
  ADI   0.003 mg/kg body weight | (1981)
Residue   Edifenphos.

Commodity code No.	Name	MRL (mg/kg)	
MM 0812	Cattle meat	0.02	(*)
MO 0812	Cattle, Edible offal of	0.02	(*)
PE 0112	Eggs	0.01	(*)
ML 0106	Milks	0.01	(*)
PM 0110	Poultry meat	0.02	(*)
PO 0111	Poultry, Edible offal of	0.02	(*)
CM 1206	Rice bran, unprocessed	1	
CM 0649	Rice, husked	0.1	
CM 1205	Rice, polished	0.02	

(100) **METHAMIDOPHOS**

JMPR 76, 79, 81, 82, 84, 85, 89, 90
ADI 0.004 mg/kg body weight | (1990)
Residue Methamidophos.
Note Methamidophos is a metabolite of acephate for which separate MRLs
are recommended (See Item 95).

Commodity code No.		Name	MRL (mg/kg)	
AL	1021	Alfalfa forage (green)	2	1/
VB	0402	Brussels sprouts	1	
MF	0812	Cattle fat	0.01	(*)
MM	0812	Cattle meat	0.01	(*)
VC	0424	Cucumber	1	
MF	0814	Goat fat	0.01	(*)
MM	0814	Goat meat	0.01	(*)
DH	1100	Hops, dry	5	
VL	0482	Lettuce, Head	1	
ML	0106	Milks	0.01	(*)
SO	0495	Rape seed	0.1	
MF	0822	Sheep fat	0.01	(*)
MM	0822	Sheep meat	0.01	(*)
VD	0541	Soya bean (dry)	0.05	1/
VR	0596	Sugar beet	0.05	
AV	0596	Sugar beet leaves or tops	1	
FT	0312	Tree tomato	0.01	(*)1/

1/ Based on treatment with acephate.

(101) **PIRIMICARB**

JMPR 76, 78, 79, 81, 82, 85,
ADI 0.02 mg/kg body weight | (1982)
Residue Sum of pirimicarb, demethyl-pirimicarb and N-formyl-(methylamino) analogue (dimethyl-formamido-pirimicarb).

Commodity code No.		Name	MRL (mg/kg)	
AL	1020	Alfalfa fodder	20	dry wt
AL	1021	Alfalfa forage (green)	50	dry wt
GC	0640	Barley	0.05	(*)
VP	0062	Beans, shelled	0.1	
VR	0574	Beetroot	0.05	(*)
VB	0400	Broccoli	1	
VB	0402	Brussels sprouts	1	
VB	0041	Cabbages, Head	1	
VB	0404	Cauliflower	1	
VS	0624	Celery	1	
FC	0001	Citrus fruits	0.05	(*) 1/
VP	0526	Common bean (pods and/ or immature seeds)	1	
SO	0691	Cotton seed	0.05	(*)
VC	0424	Cucumber	1	
FB	0278	Currant, Black	0.5	
VO	0440	Egg plant	1	
PE	0112	Eggs	0.05	(*)
VL	0476	Endive	1	
VC	0425	Gherkin	1	
VB	0405	Kohlrabi	0.5	
VA	0384	Leek	0.5	
VL	0482	Lettuce, Head	1	
MM	0095	Meat	0.05	(*)
ML	0106	Milks	0.05	(*)
GC	0647	Oats	0.05	(*)
VA	0385	Onion, Bulb	0.5	
FC	0004	Oranges, Sweet, Sour	0.5	
HH	0740	Parsley	1	
VR	0588	Parsnip	0.05	(*)
FS	0247	Peach	0.5	
VP	0063	Peas	0.2	
TN	0672	Pecan	0.05	(*)
VO	0444	Peppers, Chili	2	
VO	0445	Peppers, Sweet	1	
FS	0014	Plums (including Prunes)	0.5	
FP	0009	Pome fruits	1	

(101) PIRIMICARB

JMPR   76, 78, 79, 81, 82, 85,
ADI   0.02 mg/kg body weight | (1982)
Residue   Sum of pirimicarb, demethyl-pirimicarb and N-formyl-(methylamino)
analogue (dimethyl-formamido-pirimicarb).

Commodity code No.	Name	MRL (mg/kg)	
VR 0589	Potato	0.05	(*)
VR 0494	Radish	0.05	(*)
SO 0495	Rape seed	0.2	
FB 0272	Raspberries, Red, Black	0.5	
VL 0502	Spinach	1	
FB 0275	Strawberry	0.5	
VR 0596	Sugar beet	0.05	(*)
VO 0447	Sweet corn (corn-on-the-cob)	0.05	(*)
VO 0448	Tomato	1	
VR 0506	Turnip, Garden	0.05	(*)
VL 0473	Watercress	1	
GC 0654	Wheat	0.05	(*)

1/ (Except Orange)

(102) **MALEIC HYDRAZIDE**

JMPR   76, 77, 80, 84
ADI   5 mg/kg body weight | (1984)
Residue Sum of free and conjugated maleic hydrazide expressed as maleic
hydrazide.
Note   Based on Na or K salt, 99,9% pure and containing not more than
1 mg hydrazine/kg.

Commodity code No.	Name	MRL (mg/kg)
VA 0385	Onion, Bulb	15
VR 0589	Potato	50

(103) **PHOSMET**

JMPR 76, 77(corr. to 76 eval.), 78, 79, 81, 84, 85, 86, 87, 88
ADI 0.02 mg/kg body weight | (1979)
Residue Sum of phosmet and its oxygen analogue (fat soluble residue).
Note Scheduled for periodic review (24.241).

Commodity code No.		N a m e	MRL (mg/kg)	
AL	1020	Alfalfa fodder	40	
AL	1021	Alfalfa forage (green)	40	fresh wt
FP	0226	Apple	10	
FS	0240	Apricot	5	
FB	0020	Blueberries	10	
MM	0812	Cattle meat	1	(fat) V
FC	0001	Citrus fruits	5	
FI	0335	Feijoa	2	
FB	0269	Grapes	10	
FI	0341	Kiwifruit	15	
GC	0645	Maize	0.05	1/
AS	0645	Maize fodder	10	
AF	0645	Maize forage	10	
ML	0106	Milks	0.02	(*) F V
FS	0245	Nectarine	5	
AL	0072	Pea hay or Pea fodder (dry)	10	
AL	0528	Pea vines (green)	10	fresh wt
FS	0247	Peach	10	
FP	0230	Pear	10	
VP	0063	Peas	0.2	
VD	0072	Peas (dry)	0.02	(*)
VR	0589	Potato	0.05	
VO	0447	Sweet corn (corn-on-the-cob)	0.05	
VR	0508	Sweet potato	10	Po 2/
TN	0085	Tree nuts	0.1	

1/ Kernels and cob with husks removed
2/ Washed before analysis

(105) **DITHIOCARBAMATES**

JMPR   65, 67, 70, 74, 77, 80, 83, 84, 85, 87, 88
ADI | 2/
Note   The MRLs are determined and expressed as mg CS2/kg and refer
        to the total residues arising from use of any or each of the groups
        of dithiocarbamates. 3/

Commodity code No.	Name	MRL (mg/kg)
FP 0226	Apple	3
FI 0327	Banana	1
VR 0577	Carrot	0.5
VS 0624	Celery	5
FS 0013	Cherries	1
VP 0526	Common bean (pods and/ or immature seeds)	0.5
VC 0424	Cucumber	0.5
FB 0021	Currants, Black, Red, White	5
VL 0476	Endive	1
FB 0269	Grapes	5
VL 0482	Lettuce, Head	5
VC 0046	Melons, except Watermelon	1
FS 0247	Peach	3
FP 0230	Pear	3
FS 0014	Plums (including Prunes)	1
VR 0589	Potato	0.1
FB 0275	Strawberry	3
VO 0448	Tomato	3
GC 0654	Wheat	0.2

1/ The 1980 JMPR required further information on use patterns and data
   from residues trials before its estimates of TMRLs could be confirmed. As these requirements have not been met the proposed limits,
   which have now been adopted as CXLs, should be regarded as temporary with the exception of lettuce, head (20.104: 22.149).

2/ Ferbam, ziram: ADI 0.02 mg/kg bw. thiram: TADI 0.005, withdrawn by
   1985 JMPR. mancozeb, maneb, zineb (including zineb derived from
   nabam plus zinc sulphate): ADI 0.05 mg/kg bw, not more than 0.002
   mg/kg bw (present as ETU - 1980). Propineb: TADI 0.005 mg/kg bw,
   withdrawn by 1985 JMPR.

3/ (a) Dimethyldithiocarbamates resulting from the use of ferbam or
       ziram

   (b) Ethylenebisdithiocarbamates resulting from the use of mancozeb,
       maneb or zineb (including zineb derived from nabam plus zinc
       sulphate).

(107) **ETHIOFENCARB**

JMPR   77, 78, 81, 82, 83
ADI   0.1 mg/kg body weight | (1982)
Residue   Sum of ethiofencarb, its sulphoxide and sulphone, expressed as ethiofencarb.

Commodity code No.	Name	MRL (mg/kg)	
FP 0226	Apple	5	
FS 0240	Apricot	5	
VS 0620	Artichoke, Globe	5	
GC 0640	Barley	0.05	
VP 0062	Beans, shelled	0.2	
VB 0040	Brassica vegetables	2	
MM 0812	Cattle meat	0.02	(*)
FS 0013	Cherries	10	
VL 0466	Chinese cabbage type "pak-choi"	2	
VL 0467	Chinese cabbage, type "Pe-tsai"	5	
VP 0526	Common bean (pods and/ or immature seeds)	2	
VC 0424	Cucumber	1	
FB 0021	Currants, Black, Red, White	2	
VO 0440	Egg plant	2	
PE 0112	Eggs	0.02	(*)
AM 1051	Fodder beet	0.1	
VL 0482	Lettuce, Head	10	
ML 0106	Milks	0.02	(*)
GC 0647	Oats	0.05	
FS 0247	Peach	5	
FP 0230	Pear	5	
MM 0818	Pig meat	0.02	(*)
FS 0014	Plums (including Prunes)	5	
VR 0589	Potato	0.5	
PM 0110	Poultry meat	0.02	(*)
VR 0494	Radish	0.5	
GC 0650	Rye	0.05	
VP 0541	Soya bean (immature seeds)	0.2	
VR 0596	Sugar beet	0.1	
AV 0596	Sugar beet leaves or tops	5	
GC 0654	Wheat	0.05	
AS 0654	Wheat straw and fodder, dry	2	

(109)   **FENBUTATIN OXIDE**

JMPR   77, 79
ADI   0.03 mg/kg body weight | (1977)
Residue   Fenbutatin oxide

Commodity code No.	Name	MRL (mg/kg)
FP 0226	Apple	5
AB 0226	Apple pomace, dry	20
FS 0013	Cherries	5
FC 0001	Citrus fruits	5
AB 0001	Citrus pulp, dry	7
VC 0424	Cucumber	1
VO 0440	Egg plant	1
VC 0425	Gherkin	1
FB 0269	Grapes	5
MO 1292	Horse kidney	0.2
MO 1293	Horse liver	0.2
MO 0098	Kidney of cattle, goats, pigs and sheep	0.2
MO 0099	Liver of cattle, goats, pigs, sheep	0.2
MM 0096	Meat of cattle, goats, horses, pigs /sheep	0.02 (*)
VC 0046	Melons, except Watermelon	1
ML 0106	Milks	0.02 (*)
FS 0247	Peach	7
FP 0230	Pear	5
VO 0445	Peppers, Sweet	1
FS 0014	Plums (including Prunes)	3
FB 0275	Strawberry	3
VO 0448	Tomato	1

(110)   **IMAZALIL**

JMPR   77, 80, 84, 85, 86, 88, 89, 91
ADI   0.03 mg/kg body weight | (1991)
Residue   Imazalil.

Commodity code No.	N a m e	MRL (mg/kg)	
FI 0327	Banana	2	Po
FC 0001	Citrus fruits	5	Po
VC 0424	Cucumber	0.5	
VC 0425	Gherkin	0.5	
FT 0307	Persimmon, Japanese	2	Po
FP 0009	Pome fruits	5	Po
VR 0589	Potato	5	Po 1/
FB 0272	Raspberries, Red, Black	2	
FB 0275	Strawberry	2	
GC 0654	Wheat	0.01	(*)
AS 0654	Wheat straw and fodder, dry	0.1	

1/ Washed before analysis

(111) **IPRODIONE**

JMPR   77, 80
ADI   0.3 mg/kg body weight | (1977)
Residue Iprodione

Commodity code No.	Name	MRL (mg/kg)	
FP 0226	Apple	10	Po
VD 0071	Beans (dry)	0.2	
VC 0424	Cucumber	5	
FB 0021	Currants, Black, Red, White	5	
VA 0381	Garlic	0.1	
FB 0269	Grapes	10	
FI 0341	Kiwifruit	5	
VL 0482	Lettuce, Head	10	
VA 0385	Onion, Bulb	0.1	
FS 0247	Peach	10	Po
FP 0230	Pear	10	Po
VO 0445	Peppers, Sweet	5	
FS 0014	Plums (including Prunes)	10	
FB 0272	Raspberries, Red, Black	5	
CM 0649	Rice, husked	3	
FB 0275	Strawberry	10	
VO 0448	Tomato	5	
VS 0469	Witloof chicory (sprouts)	1	

(112) **PHORATE**

JMPR   77, 82, 83, 84, 85, 90, 91
ADI   0.0002 mg/kg body weight | (1985)
Residue   Sum of phorate, its oxygen analogue, and their sulphoxides and
          sulphones, expressed as phorate.

Commodity code No.	Name	MRL (mg/kg)	
GC 0640	Barley	0.05	
VP 0526	Common bean (pods and/ or immature seeds)	0.1	
SO 0691	Cotton seed	0.05	
PE 0112	Eggs	0.05	(*)
AM 1051	Fodder beet	0.05	
AS 0645	Maize fodder	0.2	fresh wt
MM 0095	Meat	0.05	(*)
ML 0106	Milks	0.05	(*)
SO 0495	Rape seed	0.1	
GC 0651	Sorghum	0.05	
VD 0541	Soya bean (dry)	0.05	
VR 0596	Sugar beet	0.05	
AV 0596	Sugar beet leaves or tops	1	
VO 0448	Tomato	0.1	
GC 0654	Wheat	0.05	

(113) **PROPARGITE**

JMPR   77, 78, 79, 80, 82
  ADI   0.15 mg/kg body weight | (1982)
Residue   Propargite (fat soluble).

Commodity code No.		Name	MRL (mg/kg)	
AL	1020	Alfalfa fodder	75	
AL	1021	Alfalfa forage (green)	50	
TN	0660	Almonds	0.1	(*)
FP	0226	Apple	5	
AB	0226	Apple pomace, dry	80	
FS	0240	Apricot	7	
VD	0071	Beans (dry)	0.2	
FC	0001	Citrus fruits	5	
AB	0001	Citrus pulp, dry	40	
VP	0526	Common bean (pods and/ or immature seeds)	20	
SO	0691	Cotton seed	0.1	(*)
FB	0265	Cranberry	10	
VC	0424	Cucumber	0.5	
DF	0269	Dried grapes	10	
PE	0112	Eggs	0.1	
FT	0297	Fig	2	
AB	0269	Grape pomace, dry	40	
FB	0269	Grapes	10	
DH	1100	Hops, dry	30	
GC	0645	Maize	0.1	(*)
AS	0645	Maize fodder	10	
AF	0645	Maize forage	10	
MM	0095	Meat	0.1	(fat)
ML	0106	Milks	0.1	F
AM	0738	Mint hay	50	
FS	0245	Nectarine	7	
FS	0247	Peach	7	
SO	0697	Peanut	0.1	(*)
AL	0697	Peanut fodder	10	
AL	1270	Peanut forage (green)	10	fresh wt
FP	0230	Pear	5	
FS	0014	Plums (including Prunes)	7	
VR	0589	Potato	0.1	(*)
PM	0110	Poultry meat	0.1	(fat)
GC	0651	Sorghum	5	
AF	0651	Sorghum forage (green)	10	fresh wt
AS	0651	Sorghum straw and fodder, dry	10	

(113) **PROPARGITE**

JMPR   77, 78, 79, 80, 82
  ADI   0.15 mg/kg body weight | (1982)
Residue   Propargite (fat soluble).

Commodity code No.	Name	MRL (mg/kg)
FB 0275	Strawberry	7
DT 1114	Tea, Green, Black	10
VO 0448	Tomato	2
TN 0678	Walnuts	0.1 (*)

----

(114) **GUAZATINE**

JMPR   78, 80
ADI   0.03 mg/kg body weight | (1978)
Residue Guazatine.
Note   Scheduled for periodic review (24.241).

Commodity code No.	Name	MRL (mg/kg)	
GC 0080	Cereal grains	0.1	(*)
FC 0001	Citrus fruits	5	Po
VC 0046	Melons, except Watermelon	5	Po
FI 0353	Pineapple	0.1	(*)
VR 0589	Potato	0.1	(*)
GS 0659	Sugar cane	0.1	(*)

(115)   **TECNAZENE**

JMPR   74, 78, 81, 83, 87, 89
  ADI   0.01 mg/kg body weight | (1983)
Residue   Tecnazene.

	Commodity		MRL (mg/kg)	
code No.		N a m e		
VL 0482		Lettuce, Head	2	
VR 0589		Potato	1	Po 1/
VS 0469		Witloof chicory (sprouts)	0.2	

1/ Washed before analysis

(116)   **TRIFORINE**

JMPR   77, 78
ADI   0.02 mg/kg body weight | (1978)
Residue   Determined as chloral hydrate and expressed as triforine.
Note   Scheduled for periodic review (24.241).

code No.	Commodity Name	MRL (mg/kg)	
FP 0226	Apple	2	
FB 0020	Blueberries	1	
VB 0402	Brussels sprouts	0.2	
GC 0080	Cereal grains	0.1	
FS 0013	Cherries	2	
VP 0526	Common bean (pods and/ or immature seeds)	1	
FB 0021	Currants, Black, Red, White	1	
VC 0045	Fruiting vegetables, Cucurbits	0.5	
FB 0268	Gooseberry	1	
FS 0247	Peach	5	Po
FS 0014	Plums (including Prunes)	2	
FB 0275	Strawberry	1	
VO 0448	Tomato	0.5	
FT 0312	Tree tomato	0.02	

(117)   **ALDICARB**

JMPR   79, 82, 85, 88, 90
 ADI   0.005 mg/kg body weight | (1982)
Residue   Sum of aldicarb, its sulphoxide and its sulphone, expressed as
          aldicarb.

	Commodity		MRL (mg/kg)	
code No.		N a m e		
FI	0327	Banana	0.5	
VD	0071	Beans (dry)	0.1	
FC	0001	Citrus fruits	0.2	
SB	0716	Coffee beans	0.1	
SO	0691	Cotton seed	0.1	
GC	0645	Maize	0.05	
AF	0645	Maize forage	5	fresh wt
MM	0095	Meat	0.01	(*)
ML	0106	Milks	0.01	(*)
VA	0385	Onion, Bulb	0.05	(*)
SO	0697	Peanut	0.05	(*)
TN	0672	Pecan	0.5	
VR	0589	Potato	0.5	
GC	0651	Sorghum	0.2	
AS	0651	Sorghum straw and fodder, dry	0.5	
VD	0541	Soya bean (dry)	0.02	(*)
VR	0596	Sugar beet	0.05	(*)
AV	0596	Sugar beet leaves or tops	1	
VR	0508	Sweet potato	0.1	

(118)   **CYPERMETHRIN**

JMPR   79, 81, 82, 83, 84, 85, 86, 87(corr. to 86 eval.), 88, 90
ADI   0.05 mg/kg body weight | (1981)
Residue   Cypermethrin (sum of isomers) (fat soluble).

Commodity code No.		Name	MRL (mg/kg)	
AL	1021	Alfalfa forage (green)	5	dry wt
GC	0640	Barley	0.5	
VP	0062	Beans, shelled	0.05	(*)
FB	0018	Berries and other small fruits	0.5	
VB	0040	Brassica vegetables	1	
FS	0013	Cherries	1	
FC	0001	Citrus fruits	2	
SB	0716	Coffee beans	0.05	(*)
VP	0526	Common bean (pods and/ or immature seeds)	0.5	
VC	0424	Cucumber	0.2	
MO	0105	Edible offal (Mammalian)	0.05	(*) V
VO	0440	Egg plant	0.2	
PE	0112	Eggs	0.05	(*)
VL	0480	Kale	1	
VA	0384	Leek	0.5	
VL	0482	Lettuce, Head	2	
GC	0645	Maize	0.05	(*)
AS	0645	Maize fodder	5	dry wt
MM	0095	Meat	0.2	(fat) V
ML	0106	Milks	0.05	F V
VO	0450	Mushrooms	0.05	(*)
FS	0245	Nectarine	2	
SO	0089	Oilseed, except peanut	0.2	
VA	0385	Onion, Bulb	0.1	
FS	0247	Peach	2	
SO	0697	Peanut	0.05	(*)
VP	0063	Peas	0.05	(*)
VO	0051	Peppers	0.5	
FS	0014	Plums (including Prunes)	1	
FP	0009	Pome fruits	2	
PM	0110	Poultry meat	0.05	(*)
VR	0075	Root and tuber vegetables	0.05	(*)
AS	0651	Sorghum straw and fodder, dry	5	
VD	0541	Soya bean (dry)	0.05	(*)
VL	0502	Spinach	2	
VO	0447	Sweet corn (corn-on-the-cob)	0.05	(*)
DT	1114	Tea, Green, Black	20	

(118) **CYPERMETHRIN**

JMPR   79, 81, 82, 83, 84, 85, 86, 87(corr. to 86 eval.), 88, 90
ADI   0.05 mg/kg body weight | (1981)
Residue   Cypermethrin (sum of isomers) (fat soluble).

Commodity code No.	Name	MRL (mg/kg)
VO 0448	Tomato	0.5
OR 0172	Vegetable oils, edible	0.5
GC 0654	Wheat	0.2
AS 0654	Wheat straw and fodder, dry	5

----

(119) FENVALERATE

JMPR   79, 81, 82, 84, 85, 86, 87, 88, 90
ADI   0.02 mg/kg body weight | (1986)
Residue   Fenvalerate (fat soluble).

Commodity code No.		Name	MRL (mg/kg)	
AL	1020	Alfalfa fodder	20	dry wt
VP	0061	Beans, except broad bean and soya bean	1	
VP	0062	Beans, shelled	0.1	
FB	0018	Berries and other small fruits	1	
VB	0400	Broccoli	2	
VB	0402	Brussels sprouts	2	
VB	0041	Cabbages, Head	3	
VB	0404	Cauliflower	2	
VS	0624	Celery	2	
GC	0080	Cereal grains	2	Po
FS	0013	Cherries	2	
VL	0466	Chinese cabbage type "pak-choi"	1	
FC	0001	Citrus fruits	2	
SO	0691	Cotton seed	0.2	
OC	0691	Cotton seed oil, crude	0.1	
OR	0691	Cotton seed oil, edible	0.1	
VC	0424	Cucumber	0.2	
MO	0105	Edible offal (Mammalian)	0.02	
VL	0480	Kale	10	
FI	0341	Kiwifruit	5	
VL	0482	Lettuce, Head	2	
MM	0095	Meat	1	(fat)
VC	0046	Melons, except Watermelon	0.2	
ML	0106	Milks	0.1	F
FS	0247	Peach	5	
SO	0703	Peanut, whole	0.1	
VP	0064	Peas, shelled	0.1	
VO	0445	Peppers, Sweet	0.5	
FP	0009	Pome fruits	2	
VR	0075	Root and tuber vegetables	0.05	
VD	0541	Soya bean (dry)	0.1	
VC	0431	Squash, Summer	0.5	
SO	0702	Sunflower seed	0.1	
VO	0447	Sweet corn (corn-on-the-cob)	0.1	
VO	0448	Tomato	1	
TN	0085	Tree nuts	0.2	
VC	0432	Watermelon	0.5	

(119) **FENVALERATE**

JMPR   79, 81, 82, 84, 85, 86, 87, 88, 90
  ADI   0.02 mg/kg body weight | (1986)
Residue   Fenvalerate (fat soluble).

Commodity

code No.		Name	MRL (mg/kg)	
CM	0654	Wheat bran, unprocessed	5	PoP
CF	1211	Wheat flour	0.2	PoP
CF	1212	Wheat wholemeal	2	PoP
VC	0433	Winter squash	0.5	

(120)   **PERMETHRIN**

JMPR   79, 80, 81, 82, 83, 84, 85, 86, 87, 88, 89, 91
ADI   0.05 mg/kg body weight │ (1987)

Residue   Permethrin (sum of isomers) (fat soluble residue).

Note   ADI applies to the nominal 40% cis-, 60% trans- and 25% cis- 75% trans-materials only.

Commodity code No.		Name	MRL (mg/kg)	
AL	1020	Alfalfa fodder	100	dry wt
TN	0660	Almonds	0.1	
AB	0226	Apple pomace, dry	50	
VS	0621	Asparagus	1	
VD	0071	Beans (dry)	0.1	
FB	0264	Blackberries	1	
VB	0400	Broccoli	2	
VB	0402	Brussels sprouts	1	
VB	0403	Cabbage, Savoy	5	
VB	0041	Cabbages, Head	5	
VR	0577	Carrot	0.1	
VB	0404	Cauliflower	0.5	
VS	0624	Celery	2	
GC	0080	Cereal grains	2	Po
VL	0467	Chinese cabbage, type "Pe-tsai"	5	
FC	0001	Citrus fruits	0.5	
SB	0716	Coffee beans	0.05	(*)
VP	0526	Common bean (pods and/ or immature seeds)	1	
SO	0691	Cotton seed	0.5	
OR	0691	Cotton seed oil, edible	0.1	
VC	0424	Cucumber	0.5	
FB	0021	Currants, Black, Red, White	2	
FB	0266	Dewberries	1	
MO	0105	Edible offal (Mammalian)	0.1	V
VO	0440	Egg plant	1	
PE	0112	Eggs	0.1	
VC	0425	Gherkin	0.5	
FB	0268	Gooseberry	2	
FB	0269	Grapes	2	
DH	1100	Hops, dry	50	
VR	0583	Horseradish	0.5	
VL	0480	Kale	5	
FI	0341	Kiwifruit	2	
VB	0405	Kohlrabi	0.1	
VA	0384	Leek	0.5	

(120)   **PERMETHRIN**

JMPR   79, 80, 81, 82, 83, 84, 85, 86, 87, 88, 89, 91
ADI   0.05 mg/kg body weight | (1987)
Residue   Permethrin (sum of isomers) (fat soluble residue).
Note   ADI applies to the nominal 40% cis-, 60% trans- and 25% cis- 75%
          trans-materials only.

Commodity code No.	Name	MRL (mg/kg)	
VL 0482	Lettuce, Head	2	
AS 0645	Maize fodder	100	dry wt
MM 0095	Meat	1	(fat) V
VC 0046	Melons, except Watermelon	0.1	
ML 0106	Milks	0.1	F
VO 0450	Mushrooms	0.1	
FT 0305	Olives	1	
SO 0697	Peanut	0.1	
VP 0064	Peas, shelled	0.1	
VO 0051	Peppers	1	
TN 0675	Pistachio nuts	0.05	(*)
FP 0009	Pome fruits	2	
VR 0589	Potato	0.05	(*)
PM 0110	Poultry meat	0.1	
VR 0591	Radish, Japanese	0.1	
SO 0495	Rape seed	0.05	(*)
FB 0272	Raspberries, Red, Black	1	
AS 0651	Sorghum straw and fodder, dry	20	
VD 0541	Soya bean (dry)	0.05	(*)
AL 0541	Soya bean fodder	50	dry wt
OC 0541	Soya bean oil, crude	0.1	
VL 0502	Spinach	2	
VA 0389	Spring onion	0.5	
VC 0431	Squash, Summer	0.5	
FS 0012	Stone fruits	2	
FB 0275	Strawberry	1	
VR 0596	Sugar beet	0.05	(*)
SO 0702	Sunflower seed	1	
OC 0702	Sunflower seed oil,crude	1	
OR 0702	Sunflower seed oil,edible	1	
VO 0447	Sweet corn (corn-on-the-cob)	0.1	
DT 1114	Tea, Green, Black	20	
VO 0448	Tomato	1	
VC 0433	Winter squash	0.5	

(121) **2,4,5-T**

JMPR   70, 79, 81
  ADI   0.03 mg/kg body weight | Based on 2,4,5-T containing not more than 0.01 mg T
Residue   2,4,5-T

Commodity code No.	Name	MRL (mg/kg)	
FP 0226	Apple	0.05	(*)
FS 0240	Apricot	0.05	(*)
GC 0640	Barley	0.05	(*)
MO 0105	Edible offal (Mammalian)	0.05	(*)
PE 0112	Eggs	0.05	(*)
MM 0095	Meat	0.05	(*)
ML 0106	Milks	0.05	(*)
GC 0647	Oats	0.05	(*)
GC 0649	Rice	0.05	(*)
GC 0650	Rye	0.05	(*)
AS 0081	Straw and fodder (dry) of cereal grains	2	
GS 0659	Sugar cane	0.05	(*)
GC 0654	Wheat	0.05	(*)

(122) **AMITRAZ**

JMPR   80, 83, 84, 85, 86, 89, 90
ADI   0.003 mg/kg body weight | (1984 and confirmed 1990)
Residue   Sum of amitraz and N-(2,4-dimethylphenyl)-N'-methylformamidine
           calculated as N-(2,4-dimethylphenyl)-N'-methylformamidine.
Note   The Committee requested Governments to report on their national
           legislation with the aim of securing international harmonization
           (24.134).

Commodity code No.		Name	MRL (mg/kg)	
MM	0812	Cattle meat	0.05	V
FS	0013	Cherries	0.5	
SO	0691	Cotton seed	0.5	
OC	0691	Cotton seed oil, crude	0.05	
VC	0424	Cucumber	0.5	
MO	0097	Edible offal of cattle, pigs and sheep	0.2	V
ML	0106	Milks	0.01	(*) V
FC	0004	Oranges, Sweet, Sour	0.5	
FS	0247	Peach	0.5	
MM	0818	Pig meat	0.05	V
FP	0009	Pome fruits	0.5	
MM	0822	Sheep meat	0.1	V
VO	0448	Tomato	0.5	

(123) ETRIMFOS

JMPR   80, 82, 86, 87, 88, 89, 90
ADI   0.003 mg/kg body weight | (1982)
Residue   Sum of etrimfos and its oxygen analogue (fat soluble).

Commodity code No.	Name		MRL (mg/kg)	
FP	0226	Apple	1	
FS	0240	Apricot	0.05	
VS	0620	Artichoke, Globe	0.2	
GC	0640	Barley	5	Po
VB	0402	Brussels sprouts	0.05	
VB	0041	Cabbages, Head	0.1	
MM	0812	Cattle meat	0.01	(*)
MO	0812	Cattle, Edible offal of	0.01	(*)
VB	0404	Cauliflower	0.05	
FS	0013	Cherries	0.01	(*)
VL	0467	Chinese cabbage, type "Pe-tsai"	0.1	
VP	0526	Common bean (pods and/ or immature seeds)	0.2	
VC	0424	Cucumber	0.1	
PE	0112	Eggs	0.01	(*)
FB	0269	Grapes	0.2	
VL	0480	Kale	0.5	
VB	0405	Kohlrabi	0.01	(*)
VA	0384	Leek	0.1	
GC	0645	Maize	5	Po
ML	0106	Milks	0.01	(*) F
VA	0385	Onion, Bulb	0.1	
FS	0247	Peach	0.05	
VP	0063	Peas	0.2	
FS	0014	Plums (including Prunes)	0.2	
VR	0589	Potato	0.1	
PM	0110	Poultry meat	0.02	(*)
VR	0494	Radish	0.1	
SO	0495	Rape seed	10	Po
OR	0495	Rapeseed oil, edible	0.5	
GC	0649	Rice	0.1	
VP	0541	Soya bean (immature seeds)	0.01	(*)
VR	0596	Sugar beet	0.01	(*)
AV	0596	Sugar beet leaves or tops	0.01	(*)
VO	0448	Tomato	0.2	
GC	0654	Wheat	5	Po
CM	0654	Wheat bran, unprocessed	10	PoP
CF	1211	Wheat flour	1	PoP

(123) **ETRIMFOS**

JMPR   80, 82, 86, 87, 88, 89, 90
ADI   0.003 mg/kg body weight | (1982)
Residue   Sum of etrimfos and its oxygen analogue (fat soluble).

Commodity code No.	Name	MRL (mg/kg)	
CF 1212	Wheat wholemeal	5	PoP

(124)   **MECARBAM**

JMPR   80, 83, 85, 86, 87
ADI   0.002 mg/kg body weight | (1986)
Residue   Mecarbam.

Commodity code No.	Name	MRL (mg/kg)
MM 0812	Cattle meat	0.01 (*)
ML 0812	Cattle milk	0.01
MO 0812	Cattle, Edible offal of	0.01 (*)
FC 0001	Citrus fruits	2

(126)   **OXAMYL**

JMPR   80, 83, 84, 85, 86
ADI   0.03 mg/kg body weight | (1984)
Residue   Sum of oxamyl and 2-hydroxyimino-N,N-dimethyl-2- (methylthio)
acetamide ("oxamyl oxime") expressed as oxamyl.

code No.	Commodity Name	MRL (mg/kg)	
FP 0226	Apple	2	
FI 0327	Banana	0.2	
VP 0061	Beans, except broad bean and soya bean	0.2	
VS 0624	Celery	5	
FC 0001	Citrus fruits	5	
SB 0716	Coffee beans	0.1	
SO 0691	Cotton seed	0.2	
VC 0424	Cucumber	2	
GC 0645	Maize	0.05	(*)
VC 0046	Melons, except Watermelon	2	
VA 0385	Onion, Bulb	0.05	(*)
SO 0697	Peanut	0.1	
AL 0697	Peanut fodder	2	
VO 0445	Peppers, Sweet	2	
FI 0353	Pineapple	1	
VR 0075	Root and tuber vegetables	0.1	
VD 0541	Soya bean (dry)	0.1	
VC 0431	Squash, Summer	2	
GS 0659	Sugar cane	0.05	(*)
VO 0448	Tomato	2	
VC 0432	Watermelon	2	

(127) **PHENOTHRIN**

JMPR   79, 80, 82, 84, 87, 88
ADI   0.07 mg/kg body weight | for "d-phenothrin" (1988). See 21.171
Residue   Phenothrin (sum of (+)-trans- and (+)-cis- isomers).

Commodity code No.	Name	MRL (mg/kg)	
GC 0640	Barley	2	Po
CM 0649	Rice, husked	0.1	
GC 0651	Sorghum	2	Po
GC 0654	Wheat	2	Po
CM 0654	Wheat bran, unprocessed	5	PoP
CF 1211	Wheat flour	1	PoP
CF 1210	Wheat germ	5	PoP
CF 1212	Wheat wholemeal	2	PoP

(128) **PHENTHOATE**

JMPR   80, 81, 84
 ADI   0.003 mg/kg body weight | (1984)
Residue   Phenthoate (fat soluble).

code No.		Commodity N a m e	MRL (mg/kg)	
MM	0812	Cattle meat	0.05	(*)
FC	0001	Citrus fruits	1	
PE	0112	Eggs	0.05	(*)
ML	0106	Milks	0.01	(*) F
CM	0649	Rice, husked	0.05	

(130) **DIFLUBENZURON**

JMPR   81, 83, 84, 85, 88
ADI   0.02 mg/kg body weight | (1985)
Residue   Diflubenzuron.

Commodity code No.	Name	MRL (mg/kg)	
FP 0226	Apple	1	
VB 0402	Brussels sprouts	1	
VB 0041	Cabbages, Head	1	
FC 0001	Citrus fruits	1	
SO 0691	Cotton seed	0.2	
MO 0105	Edible offal (Mammalian)	0.05	(*)
PE 0112	Eggs	0.05	(*)
MM 0095	Meat	0.05	(*)
ML 0106	Milks	0.05	(*)
VO 0450	Mushrooms	0.1	
FP 0230	Pear	1	
FS 0014	Plums (including Prunes)	1	
PM 0110	Poultry meat	0.05	(*)
VD 0541	Soya bean (dry)	0.1	
VO 0448	Tomato	1	

(131) **ISOFENPHOS**

JMPR   81, 82, 84, 85, 86, 88
ADI   0.001 mg/kg body weight | (1986)
Residue   Sum of isofenphos and its oxygen analogue (fat soluble).

Commodity code No.		Name	MRL (mg/kg)	
FI	0327	Banana	0.02	(*)
VB	0040	Brassica vegetables	0.1	
VR	0578	Celeriac	0.02	(*)
VS	0624	Celery	0.02	(*)
FC	0001	Citrus fruits	2	
MO	0105	Edible offal (Mammalian)	0.02	(*)
GC	0645	Maize	0.02	(*)
AS	0645	Maize fodder	0.5	dry wt
MF	0100	Mammalian fats (except milk fats)	0.02	(*)
MM	0095	Meat	0.02	(*)
ML	0106	Milks	0.01	(*) F
VA	0385	Onion, Bulb	0.1	
VR	0589	Potato	0.1	
PF	0111	Poultry fats	0.02	(*)
PM	0110	Poultry meat	0.02	(*)
PO	0111	Poultry, Edible offal of	0.02	(*)
SO	0495	Rape seed	0.02	(*)
VR	0497	Swede	0.02	(*)
VO	0447	Sweet corn (corn-on-the-cob)	0.02	(*)
AS	0447	Sweet corn fodder	0.5	
VR	0506	Turnip, Garden	0.02	(*)

(132)   **METHIOCARB**

JMPR   81, 83, 84, 85, 86, 87, 88
ADI   0.001 mg/kg body weight | (1981, confirmed 1987)
Residue   Sum of methiocarb, its sulphoxide and its sulphone, expressed as
          methiocarb

Commodity code No.	Name	MRL (mg/kg)	
VS 0620	Artichoke, Globe	0.05	(*)
VB 0400	Broccoli	0.2	
VB 0402	Brussels sprouts	0.2	
VB 0041	Cabbages, Head	0.2	
VB 0404	Cauliflower	0.2	
GC 0080	Cereal grains	0.05	(*)
FC 0001	Citrus fruits	0.05	(*)
PE 0112	Eggs	0.05	(*)
TN 0666	Hazelnuts	0.05	(*)
VL 0482	Lettuce, Head	0.2	
VL 0483	Lettuce, Leaf	0.2	
MM 0095	Meat	0.05	(*)
ML 0106	Milks	0.05	(*)
PM 0110	Poultry meat	0.05	(*)
SO 0495	Rape seed	0.05	(*)
VR 0596	Sugar beet	0.05	(*)
VO 0447	Sweet corn (corn-on-the-cob)	0.05	(*)

(133) **TRIADIMEFON**

JMPR   79, 81, 83, 84, 85, 86, 87, 88, 89
ADI   0.03 mg/kg body weight | (1985)
Residue Sum of triadimefon and triadimenol. See 21.175
Note   24th CCPR noted that because this compound was closely related to
        triadimenol (168), a complete residue review of both compounds was
        needed by the JMPR in order to derive separate MRLs. The review will
        take place at the 1992 JMPR (24.139).

Commodity code No.	Name	MRL (mg/kg)	
VD 0524	Chick-pea (dry)	0.1	(*)
SB 0716	Coffee beans	0.1	(*)
FB 0021	Currants, Black, Red, White	1	
PE 0112	Eggs	0.1	(*)
AM 1051	Fodder beet	0.1	(*)
AV 1051	Fodder beet leaves or tops	0.1	(*)
VC 0045	Fruiting vegetables, Cucurbits	0.2	
DH 1100	Hops, dry	15	
FI 0345	Mango	0.1	(*)
MM 0095	Meat	0.1	(*)
ML 0106	Milks	0.1	(*)
VA 0387	Onion, Welsh	0.1	(*)
VP 0063	Peas	0.1	(*)
VO 0445	Peppers, Sweet	0.5	
FI 0353	Pineapple	3	Po
FP 0009	Pome fruits	0.5	
PM 0110	Poultry meat	0.1	(*)
VA 0389	Spring onion	0.1	(*)
FB 0275	Strawberry	0.2	
VR 0596	Sugar beet	0.1	(*)
AV 0596	Sugar beet leaves or tops	2	
VO 0448	Tomato	0.5	

(135) **DELTAMETHRIN**

JMPR   80, 81, 82, 84, 85, 86, 87, 88, 90
ADI   0.01 mg/kg body weight | (1982)
Residue   Deltamethrin (fat soluble).

Commodity code No.	Name	MRL (mg/kg)	
VS 0620	Artichoke, Globe	0.05	
FI 0327	Banana	0.05	
VD 0071	Beans (dry)	1	Po
VB 0040	Brassica vegetables	0.2	
VA 0036	Bulb vegetables, except Fennel, Bulb	0.1	
SB 0715	Cacao beans	0.05	
GC 0080	Cereal grains	1	Po
SB 0716	Coffee beans	2	Po
MO 0105	Edible offal (Mammalian)	0.05	V
PE 0112	Eggs	0.01	(*)
VD 0561	Field pea (dry)	1	Po
FT 0297	Fig	0.01	(*)
VO 0050	Fruiting vegetables other than Cucurbits	0.2	
VC 0045	Fruiting vegetables, Cucurbits	0.2	
FB 0269	Grapes	0.05	
DH 1100	Hops, dry	5	
FI 0341	Kiwifruit	0.05	
VL 0053	Leafy vegetables	0.5	
AL 0157	Legume animal feeds	0.5	dry wt
VP 0060	Legume vegetables	0.1	
VD 0533	Lentil (dry)	1	Po
FC 0003	Mandarins	0.05	
VC 0046	Melons, except Watermelon	0.01	(*)
ML 0106	Milks	0.02	F V
VO 0450	Mushrooms	0.01	(*)
SO 0088	Oilseed	0.1	
SO 0089	Oilseed, except peanut	0.1	
FT 0305	Olives	0.1	
FC 0004	Oranges, Sweet, Sour	0.05	
SO 0697	Peanut	0.01	(*)
FI 0353	Pineapple	0.01	(*)
FP 0009	Pome fruits	0.1	
PM 0110	Poultry meat	0.01	(*)
PO 0111	Poultry, Edible offal of	0.01	(*)
VR 0075	Root and tuber vegetables	0.01	
FS 0012	Stone fruits	0.05	
AS 0081	Straw and fodder (dry) of cereal grains	0.5	

(135)   **DELTAMETHRIN**

JMPR   80, 81, 82, 84, 85, 86, 87, 88, 90
  ADI   0.01 mg/kg body weight | (1982)
Residue   Deltamethrin (fat soluble).

Commodity		MRL (mg/kg)
code No.	N a m e	
FB 0275	Strawberry	0.05
DT 1114	Tea, Green, Black	10

(137)   **BENDIOCARB**

JMPR   82, 84, 89, 90
ADI   0.004 mg/kg body weight | (1984)
Residue   Commodities of plant origin: unconjugated bendiocarb.Commodities of a-
          nimal origin:sum of conjugated/unconjugated bendiocarb,2,2-dimethyl-1,
          3-benzodioxol-4-ol/N-hydroxymethyl bendiocarb expressed as bendiocarb.

Commodity code No.	Name	MRL (mg/kg)		
VR 0574	Beetroot	0.05	(*)	
MF 0812	Cattle fat	0.05	(*)	V
MM 0812	Cattle meat	0.05	(*)	V
MO 0812	Cattle, Edible offal of	0.05	(*)	V 1/
MO 1280	Cattle, kidney	0.2	(*)	V
PE 0112	Eggs	0.05	(*)	
GC 0645	Maize	0.05	(*)	
AS 0645	Maize fodder	0.05	(*)	
AF 0645	Maize forage	0.05	(*)	
ML 0106	Milks	0.05	(*)	V
VO 0450	Mushrooms	0.1		T
VR 0589	Potato	0.05	(*)	
PF 0111	Poultry fats	0.05	(*)	
PM 0110	Poultry meat	0.05	(*)	
PO 0111	Poultry, Edible offal of	0.05	(*)	
AS 0649	Rice straw and fodder, dry	1		T
CM 0649	Rice, husked	0.02	(*)	T
VR 0596	Sugar beet	0.05	(*)	
AV 0596	Sugar beet leaves or tops	0.05	(*)	

1/ (Except kidney)

(138)   **METALAXYL**

JMPR   82, 84, 85, 86, 87, 89, 90
ADI   0.03 mg/kg body weight | (1982)
Residue   Metalaxyl (See 22.185).

Commodity code No.	Name	MRL (mg/kg)		
FP 0226	Apple	0.05	(*)	1/
VS 0621	Asparagus	0.05	(*)	
FI 0326	Avocado	0.2		
VB 0402	Brussels sprouts	0.2		
SB 0715	Cacao beans	0.2		
VR 0577	Carrot	0.05	(*)	
GC 0080	Cereal grains	0.05	(*)	
FC 0001	Citrus fruits	5	Po	
SO 0691	Cotton seed	0.05	(*)	
VC 0424	Cucumber	0.5		
VC 0425	Gherkin	0.5		
FB 0269	Grapes	1		
DH 1100	Hops, dry	10		
VC 0046	Melons, except Watermelon	0.2		
SO 0697	Peanut	0.1		
VP 0064	Peas, shelled	0.05	(*)	
VO 0051	Peppers	1		
VR 0589	Potato	0.05	(*)	
FB 0272	Raspberries, Red, Black	0.2		
VD 0541	Soya bean (dry)	0.05	(*)	
VC 0431	Squash, Summer	0.2		
VR 0596	Sugar beet	0.05	(*)	
SO 0702	Sunflower seed	0.05	(*)	
VO 0448	Tomato	0.5		
VC 0432	Watermelon	0.2		
VC 0433	Winter squash	0.2		

1/ Recommended deletion following the adoption of MRL for pome fruits
   by the Commission.

(141) **PHOXIM**

JMPR   82, 83, 84, 86, 87, 88
ADI   0.001 mg/kg body weight | (1984)
Residue   Phoxim (fat soluble residue)

Commodity code No.	Name	MRL (mg/kg)	
VB 0403	Cabbage, Savoy	0.05	(*)
MM 0812	Cattle meat	0.2	(fat) V
VB 0404	Cauliflower	0.05	(*)
GC 0080	Cereal grains	0.05	(*)
VP 0526	Common bean (pods and/ or immature seeds)	0.05	(*)
SO 0691	Cotton seed	0.05	(*)
VL 0482	Lettuce, Head	0.1	
ML 0106	Milks	0.05	F V
VA 0385	Onion, Bulb	0.05	(*)
VR 0589	Potato	0.05	(*)
MM 0822	Sheep meat	0.5	(fat) V
VO 0447	Sweet corn (corn-on-the-cob)	0.05	(*)
VO 0448	Tomato	0.2	

(142) **PROCHLORAZ**

JMPR 83, 85, 87, 88, 89, 90
ADI 0.01 mg/kg body weight | (1983)
Residue Sum of prochoraz and its metabolites containing the 2,4,6-trichloro-
phenol moiety, expressed as prochloraz.
Note MRLs cover cumulative residues from pre- and post-harvest treatments.
The Committee noted concerns about the dose level fed to animals in
transfer studies and about a different limit of determination used
in residue studies (24.146).

Commodity code No.		N a m e	MRL (mg/kg)	
FI	0326	Avocado	5	Po
FI	0327	Banana	5	Po
GC	0640	Barley	0.5	
AS	0640	Barley straw and fodder, dry	15	
SB	0716	Coffee beans	0.2	
FI	0345	Mango	2	Po
VO	0450	Mushrooms	2	
AS	0647	Oat straw and fodder, dry	15	
GC	0647	Oats	0.5	
FC	0004	Oranges, Sweet, Sour	5	Po
FI	0350	Papaya	1	Po
SO	0495	Rape seed	0.5	
GC	0650	Rye	0.5	
AS	0650	Rye straw and fodder, dry	15	
FS	0012	Stone fruits	0.05	
GC	0654	Wheat	0.5	
AS	0654	Wheat straw and fodder, dry	15	

(144) **BITERTANOL**

JMPR   83, 84, 86, 87, 88, 89
ADI   0.01 mg/kg body weight | (1988)
Residue   Bitertanol

Commodity code No.	Name	MRL (mg/kg)	
FI 0327	Banana	0.5	
AL 1030	Bean forage (green)	10	
FS 0013	Cherries	2	
VP 0526	Common bean (pods and/ or immature seeds)	0.5	
VC 0424	Cucumber	0.5	
AF 0647	Oat forage (green)	0.1	(*)
AS 0647	Oat straw and fodder, dry	0.1	(*)
GC 0647	Oats	0.1	(*)
SO 0697	Peanut	0.1	(*)
AL 1270	Peanut forage (green)	20	
FS 0014	Plums (including Prunes)	2	
FP 0009	Pome fruits	2	
GC 0650	Rye	0.1	(*)
AF 0650	Rye forage (green)	0.1	(*) fresh wt
AS 0650	Rye straw and fodder, dry	0.1	(*)
GC 0654	Wheat	0.1	(*)
AS 0654	Wheat straw and fodder, dry	0.1	(*)

(146) **CYHALOTHRIN**

JMPR   84, 86, 88
  ADI   0.02 mg/kg body weight | (1984)
Residue   Cyhalothrin (sum of all isomers).

Commodity code No.	Name	MRL (mg/kg)	
VB 0041	Cabbages, Head	0.2	
SO 0691	Cotton seed	0.02	(*)
OC 0691	Cotton seed oil, crude	0.02	(*)
OR 0691	Cotton seed oil, edible	0.02	(*)
FP 0009	Pome fruits	0.2	
VR 0589	Potato	0.02	(*)

(147) **METHOPRENE**

JMPR   84, 86, 87, 88, 89
ADI   0.1 mg/kg body weight | (1987)
Residue   Methoprene (fat soluble residue).

Commodity code No.		Name	MRL (mg/kg)	
ML	0812	Cattle milk	0.05	F V
GC	0080	Cereal grains	5	Po
MO	0105	Edible offal (Mammalian)	0.1	
PE	0112	Eggs	0.05	
OR	0645	Maize oil, edible	0.2	(*) PoP
MM	0095	Meat	0.2	(fat) V
VO	0450	Mushrooms	0.2	
SO	0697	Peanut	2	
CM	0654	Wheat bran, unprocessed	10	PoP
CF	1211	Wheat flour	2	PoP
CF	1212	Wheat wholemeal	5	PoP

----

(148) **PROPAMOCARB**

JMPR   84, 86, 87
  ADI   0.1 mg/kg body weight | (1986)
Residue   Propamocarb (base)

Commodity code No.	Name	MRL (mg/kg)
VR 0574	Beetroot	0.2
VB 0402	Brussels sprouts	1
VB 0041	Cabbages, Head	0.1
VB 0404	Cauliflower	0.2
VS 0624	Celery	0.2
VC 0424	Cucumber	2
VL 0482	Lettuce, Head	10
VO 0445	Peppers, Sweet	1
VR 0494	Radish	5
FB 0275	Strawberry	0.1
VO 0448	Tomato	1

(149)   **ETHOPROPHOS**

JMPR   83, 84, 87
ADI   0.0003 mg/kg body weight | (1987)
Residue   Ethoprophos.

Commodity code No.	Name	MRL (mg/kg)	
FI 0327	Banana	0.02	(*)
VR 0574	Beetroot	0.02	(*)
VB 0041	Cabbages, Head	0.02	(*)
VC 0424	Cucumber	0.02	(*)
VC 0425	Gherkin	0.02	(*)
FB 0269	Grapes	0.02	(*)
VL 0482	Lettuce, Head	0.02	(*)
GC 0645	Maize	0.02	(*)
AS 0645	Maize fodder	0.02	(*)
AF 0645	Maize forage	0.02	(*)
VC 0046	Melons, except Watermelon	0.02	(*)
VA 0385	Onion, Bulb	0.02	(*)
SO 0697	Peanut	0.02	(*)
AL 0697	Peanut fodder	0.02	(*)
VP 0063	Peas	0.02	(*)
VO 0051	Peppers	0.02	(*)
FI 0353	Pineapple	0.02	(*)
AM 0353	Pineapple fodder	0.02	(*)
AV 0353	Pineapple forage	0.02	(*)
VR 0589	Potato	0.02	(*)
VD 0541	Soya bean (dry)	0.02	(*)
AL 0541	Soya bean fodder	0.02	(*)
FB 0275	Strawberry	0.02	(*)
GS 0659	Sugar cane	0.02	(*)
AM 0659	Sugar cane fodder	0.02	(*)
AV 0659	Sugar cane forage	0.02	(*)
VR 0508	Sweet potato	0.02	(*)
VO 0448	Tomato	0.02	(*)
VR 0506	Turnip, Garden	0.02	(*)

(151)   **DIMETHIPIN**

JMPR   85, 87, 88
ADI   0.02 mg/kg body weight | (1988)
Residue   Dimethipin.

	Commodity		
code No.	N a m e	MRL (mg/kg)	
SO	0691	Cotton seed	0.5
OC	0691	Cotton seed oil, crude	0.1
OR	0691	Cotton seed oil, edible	0.02 (*)
MO	0105	Edible offal (Mammalian)	0.02 (*)
PE	0112	Eggs	0.02 (*)
SO	0693	Linseed	0.2
MM	0095	Meat	0.02 (*)
ML	0106	Milks	0.02 (*)
VR	0589	Potato	0.05 (*)
PM	0110	Poultry meat	0.02 (*)
PO	0111	Poultry, Edible offal of	0.02 (*)
SO	0495	Rape seed	0.1
SO	0702	Sunflower seed	0.5
OC	0702	Sunflower seed oil,crude	0.1
OR	0702	Sunflower seed oil,edible	0.02 (*)

(152) **FLUCYTHRINATE**

JMPR   85, 87, 88, 89, 90
ADI   0.02 mg/kg body weight | (1985)
Residue   Flucythrinate (fat soluble).

	Commodity		MRL (mg/kg)	
	code No.	N a m e		
VS	0620	Artichoke, Globe	0.5	
GC	0640	Barley	0.2	
AS	0640	Barley straw and fodder, dry	5	
VD	0071	Beans (dry)	0.05	(*)
VB	0041	Cabbages, Head	0.5	
SB	0716	Coffee beans	0.05	(*)
SO	0691	Cotton seed	0.1	
OC	0691	Cotton seed oil, crude	0.2	
OR	0691	Cotton seed oil, edible	0.2	
VD	0561	Field pea (dry)	0.05	(*)
VB	0042	Flowerhead brassicas	0.2	
FB	0269	Grapes	1	
DH	1100	Hops, dry	10	
GC	0645	Maize	0.05	(*)
AS	0647	Oat straw and fodder, dry	5	
GC	0647	Oats	0.2	
FS	0247	Peach	0.5	
FP	0009	Pome fruits	0.5	
VR	0589	Potato	0.05	(*)
VR	0591	Radish, Japanese	0.05	(*)
SO	0495	Rape seed	0.05	(*)
VR	0596	Sugar beet	0.05	(*)
AV	0596	Sugar beet leaves or tops	2	
VO	1275	Sweet corn (kernels)	0.05	(*)
DT	1114	Tea, Green, Black	20	
VO	0448	Tomato	0.2	
GC	0654	Wheat	0.2	
AS	0654	Wheat straw and fodder, dry	5	

(154) **THIODICARB**

JMPR   85, 86, 87, 88
ADI   0.03 mg/kg body weight | (1986)
Residue Sum of thiodicarb, methomyl and methyl hydroxythioacetimidate
        ("methomyl oxime"), expressed as thiodicarb (JMPR 88).
Note   22nd CCPR decided that entries related to thiodicarb would be
        deleted when the MRLs for methomyl (94) reach Step 8 (22.126,201).

Commodity code No.		Name	MRL (mg/kg)	
MM	0812	Cattle meat	0.02	(*)
ML	0812	Cattle milk	0.02	(*)
SO	0691	Cotton seed	0.5	
GC	0645	Maize	0.05	(*)
AS	0645	Maize fodder	50	fresh wt
AF	0645	Maize forage	50	fresh wt
VD	0541	Soya bean (dry)	0.2	
VO	0447	Sweet corn (corn-on-the-cob)	2	
VO	0448	Tomato	1	

(155)   **BENALAXYL**

JMPR   86, 87, 88
ADI   0.05 mg/kg body weight | (1987)
Residue   Benalaxyl.

Commodity code No.	Name	MRL (mg/kg)
VC 0424	Cucumber	0.05
FB 0269	Grapes	0.2
DH 1100	Hops, dry	0.2
VC 0046	Melons, except Watermelon	0.1
VA 0385	Onion, Bulb	0.2
VO 0445	Peppers, Sweet	0.05
VR 0589	Potato	0.01 (*)
VO 0448	Tomato	0.5

(156) **CLOFENTEZINE**

JMPR   86, 87, 89, 90
ADI   0.02 mg/kg body weight | (1986)
Residue   Sum of all residues containing the 2-chlorobenzoyl moiety,
expressed as clofentezine.

Commodity code No.		Name	MRL (mg/kg)	
MM	0812	Cattle meat	0.05	(*)
ML	0812	Cattle milk	0.01	(*)
MO	0812	Cattle, Edible offal of	0.1	
VC	0424	Cucumber	1	
PE	0112	Eggs	0.05	(*)
FB	0269	Grapes	0.2	
FP	0009	Pome fruits	0.5	
PM	0110	Poultry meat	0.05	(*)
PO	0111	Poultry, Edible offal of	0.05	(*)
FS	0012	Stone fruits	0.2	
FB	0275	Strawberry	2	

(158) **GLYPHOSATE**

JMPR   86, 87, 88
ADI   0.3 mg/kg body weight | (1986)
Residue   Glyphosate (See 21.218)

Commodity code No.	Name	MRL (mg/kg)	
GC 0640	Barley	20	
VD 0071	Beans (dry)	2	
MM 0812	Cattle meat	0.1	(*)
ML 0812	Cattle milk	0.1	(*)
MO 0812	Cattle, Edible offal of	2	
SO 0691	Cotton seed	0.5	
PE 0112	Eggs	0.1	(*)
AS 0162	Hay or fodder (dry) of grasses	50	
FI 0341	Kiwifruit	0.1	(*)
GC 0645	Maize	0.1	(*)
GC 0647	Oats	20	
VD 0072	Peas (dry)	5	
MM 0818	Pig meat	0.1	(*)
MO 0818	Pig, Edible offal of	1	
PM 0110	Poultry meat	0.1	(*)
SO 0495	Rape seed	10	
GC 0649	Rice	0.1	(*)
GC 0651	Sorghum	0.1	(*)
VD 0541	Soya bean (dry)	5	
VP 0541	Soya bean (immature seeds)	0.2	
AL 0541	Soya bean fodder	20	
AL 1265	Soya bean forage (green)	5	
AS 0081	Straw and fodder (dry) of cereal grains	100	
VO 0447	Sweet corn (corn-on-the-cob)	0.1	(*)
GC 0654	Wheat	5	
CF 1211	Wheat flour	0.5	
CF 1212	Wheat wholemeal	5	

(159) **VINCLOZOLIN**

JMPR   86, 87, 88, 89
ADI   0.07 mg/kg body weight | (1988)
Residue   Sum of vinclozolin and all metabolites containing the 3,5-dichloro-
          aniline moiety, expressed as vinclozolin. 20.185

Commodity code No.	Name	MRL (mg/kg)	
FB 0264	Blackberries	5	
FB 0020	Blueberries	5	
VB 0041	Cabbages, Head	1	
MM 0812	Cattle meat	0.05	(*)
ML 0812	Cattle milk	0.05	(*)
VB 0404	Cauliflower	1	
FS 0013	Cherries	5	Po
PE 0840	Chicken eggs	0.05	(*)
PM 0840	Chicken meat	0.05	(*)
VR 0469	Chicory, roots	5	
VP 0526	Common bean (pods and/ or immature seeds)	2	
VC 0424	Cucumber	1	
FB 0021	Currants, Black, Red, White	5	
FB 0266	Dewberries	5	
VP 0529	Garden pea, shelled	1	
VC 0425	Gherkin	1	
FB 0268	Gooseberry	5	
FB 0269	Grapes	5	
DH 1100	Hops, dry	40	
FI 0341	Kiwifruit	10	
VC 0046	Melons, except Watermelon	1	
VA 0385	Onion, Bulb	1	
FS 0247	Peach	5	Po
VO 0445	Peppers, Sweet	3	
FP 0009	Pome fruits	1	
VR 0589	Potato	0.1	
SO 0495	Rape seed	1	
FB 0272	Raspberries, Red, Black	5	
FB 0275	Strawberry	10	
VO 0448	Tomato	3	
VS 0469	Witloof chicory (sprouts)	2	

(160) **PROPICONAZOLE**

JMPR   87, 91
ADI   0.04 mg/kg body weight | (1987)
Residue   Propiconazole

Commodity code No.	Name	MRL (mg/kg)	
TN 0660	Almonds	0.05	
FI 0327	Banana	0.1	
SB 0716	Coffee beans	0.1	
MO 0105	Edible offal (Mammalian)	0.05	
PE 0112	Eggs	0.05	(*)
FB 0269	Grapes	0.5	
FI 0345	Mango	0.05	
MM 0095	Meat	0.05	(*)
ML 0106	Milks	0.01	(*)
SO 0697	Peanut	0.05	
SO 0703	Peanut, whole	0.1	
TN 0672	Pecan	0.05	
PM 0110	Poultry meat	0.05	(*)
SO 0495	Rape seed	0.05	
FS 0012	Stone fruits	1	
VR 0596	Sugar beet	0.05	
AV 0596	Sugar beet leaves or tops	0.1	
GS 0659	Sugar cane	0.05	

(161) **PACLOBUTRAZOL**

JMPR   88, 89
  ADI   0.1 mg/kg body weight | (1988)
Residue   Paclobutrazol.

Commodity code No.	Name	MRL (mg/kg)
FP 0226	Apple	0.5
FS 0012	Stone fruits	0.05

(162)   **TOLYLFLUANID**

JMPR   88, 90
ADI   0.1 mg/kg body weight | (1988)
Residue   Tolyfluanid.

code No.	Commodity Name	MRL (mg/kg)
FB 0021	Currants, Black, Red, White	5
VC 0425	Gherkin	2
VL 0482	Lettuce, Head	1
FP 0009	Pome fruits	5
FB 0275	Strawberry	3
VO 0448	Tomato	2

(167)   **TERBUFOS**

JMPR   89, 90
  ADI   0.0002 mg/kg body weight | (1989)
Residue   Sum of terbufos, its oxygen analogue and their sulphoxides and
          sulphones, expressed as terbufos.
Note   23rd CCPR agreed to request JMPR review of new data to be generated
        with a 0.01 mg/kg limit of determination (23.213).

Commodity code No.		Name	MRL (mg/kg)	
FI	0327	Banana	0.05	
GC	0640	Barley	0.01	(*)
ML	0812	Cattle milk	0.01	(*)
PE	0112	Eggs	0.01	(*)
AV	1051	Fodder beet leaves or tops	1	
GC	0645	Maize	0.01	(*)
AF	0645	Maize forage	1	
GC	0656	Popcorn	0.01	(*)
GC	0654	Wheat	0.01	(*)

SECTION 2

CODEX CLASSIFICATION
OF
FOODS AND ANIMAL FEEDS

FOREWORD

The Codex Classification of food and animal feed commodities moving in trade and the description of the various items and groups of food and animal feedstuffs included in the present document have been developed by the Codex Committee on Pesticide Residues. It was first adopted by the 18th Session of the Codex Alimentarius Commission, (1989).

The Codex Classification includes food commodities and animal feedstuffs for which Codex maximum residue limits will not necessarily be established. The Classification is intended to be as complete a listing of food commodities in trade as possible, classified into groups on the basis of the commodity's similar potential for pesticide residues.

The Classification may also be appropriate for other purposes such as setting maximum levels for other types of residues or for other contaminants in food. The Codex Classification should be consulted in order to obtain a precise description of the food or animal feed commodities and, especially, in cases where Codex maximum residue limits have been set for groups of food and groups of animal feedstuffs. The Codex Classification is intended to promote harmonization of the terms used to describe commodities which are subject to maximum residue limits and of the approach to grouping commodities with similar potential for residue for which a common group maximum residue limit can be set.

EXPLANATORY NOTES

Introduction

The Codex Classification of Foods and Feeds is intended primarily to ensure the use of uniform nomenclature and secondarily to classify foods into groups and/or sub-groups for the purpose of establishing group maximum residue limits for commodities with similar characteristics and residue potential.

The major differences in exposure to pesticides and metabolites of pesticides in plants and animals call for a primary classification into foods and feeds of plant origin and those of animal origin. Processed foods prepared from these primary food commodities are again separated into those of plant origin and of animal origin.

Multi-ingredient manufactured foods containing ingredients of both plant and animal origin are listed as plant or animal origin depending upon the main ingredients.

In the event that residues are greater in the processed food than in the raw agricultural commodity from which it is derived, a separate MRL should be considered for the processed food. In addition, there are a number of situations where special considerations may be needed:

(i)   when the processed food represents the sole or major food intake of infants and young children;

(ii)   when toxic interaction or degradation products from pesticides are found in the food during or after processing;

(iii)   when a significant residue results from a pesticide used in processing or storage practice (including impregnation of wrapping materials).

Crops, parts of crops and other commodities which are used as animal feed, e.g. alfalfa, sugar beet tops, pea vines and hay, may contain residues and hence give rise to residue in products of animal origin. Such commodities are listed for convenience in a separate class: Class C - Primary Animal Feed Commodities.

The food commodities selected for this classification are mainly those having current or potential significance in international or national trade. A limited number of commodities of regional importance have also been included.

Common and scientific names of commodities

A single food or animal feed is often known under several common names. In a like manner, the use of similar names for different commodities is not uncommon. In this classification uniform terminology associated with recognised scientific names has been developed.

Different common names used for the same or a closely related commodity, or cultivars of the same agricultural crop have been cross-referenced as far as possible to a single common name. Similarly, a qualification has been used to modify a common name, where the use of a generic name fails to identify the specific commodity adequately. However, specific cultivars distinguished by their own common names in agricultural practice and in international trade are indicated with these common and scientific species and cultivar names, e.g. nectarine, a specific peach variety or cultivar is distinguished from all other peach cultivars.

In cases where commonly used regional or national English names are different from the correct English common name, priority is given to the latter and the other names are referred to it e.g., "corn" is referred to the correct name, "maize".

The common names, scientific names and descriptions used in existing Codex standards were taken into account and given priority as far as possible. The terminology used in this classification closely follows "Plant and Plant Products of Economic Importance Terminology Bulletin (1974)" and the terminology used in various FAO statistics.

Within the broad classes of Foods of Plant origin and Foods of Animal Origin, usage distinguishes groups of similar commodities such as Fruits, Vegetables, Nuts, Mammalian Products, Aquatic Animal Products etc. The present classification uses 19 types of commodities:

Class	A Primary Food Commodities of Plant Origin	5 types
	B Primary Food Commodities of Animal Origin	5 types
	C Primary Feed Commodities	1 type
	D Processed Food of Plant Origin	4 types
	E Processed Food of Animal Origin	4 types

The "Type" definitions developed for this classification are based on physical characteristics and traditional use and to a lesser extent on botanical or zoological associations. A food "Type" is in general much too broad to be covered by a single (general) maximum residue limit.

Within the "Types", groups have been developed whose members show similarities in their behaviour with respect to residues and in the nature of the agricultural practices to which they are subjected and, to a certain extent, in their botanical or zoological associations.

The terminology and definitions in the classification recognize and respect, as far as possible, regional differences which occur in the use of certain food commodities. Sometimes, these are inconsistent with regard to the "Types" to which the commodity(ies) belong.

There are no universally recognized guides to distinguish between different uses of the same commodity, e.g., a herb may be used sparingly for flavouring of other food commodities in one geographical region but in substantial amounts as a pot vegetable in another region. If, however, a commodity exists in forms which differ in appearance, or if different cultivars are involved, the commodity crop or animal species) is separately included in two or more groups, e.g. turnip, listed as a root crop and turnip leaves, listed with leafy vegetables.

As far as practicable, the commodity "Group" or "Sub-Group" names currently used by the Codex Committee on Pesticide Residues for the establishment of Codex Maximum Residue Limits for pesticides have been used. However, in some cases the definition of the commodities covered by existing "Group" maximum residue limits requires clarification. For this reason, existing maximum residue limits for groups of commodities are occasionally not applicable to the groups used in this classification.

For each group in the classification, reference is made to the potential for acquiring residues, the portions of the crops or animal which are normally consumed and the portions of the commodities to which the MRLs apply (and which are analyzed).

Computerized System elaborated in D-Base IV for the Codex Classification of Foods and Animal Feeds

The Codex Classification of Foods and Animal Feeds was published in Part 4 of the "Guide to Codex Recommendations Concerning Pesticide Residues" in 1989 (CAC/PR 4-1989). The Codex Classification uses five classes and each class is divided into types. All types are divided into groups. A review of the whole system used in order to link the classification by programme to the file which contains all Codex pesticide/commodity combinations was carried out for a computerized management of the Classification.

With reference to the need to give a code numnber to all commodities for their identification by programme, a literary code has been maintained and a progressive number of the Series 4000 has been allocated. This Series refers to a large group of synonyms. Through these changes, each commodity will be represented by the following information in the computerized system: class; type; group; letter code; numerical code; description.

INDEX OF CLASSES, TYPES AND GROUPS
OF COMMODITIES

INDEX OF CLASSES, TYPES AND GROUPS OF COMMODITIES

INDEX OF CLASSES, TYPES AND GROUPS OF COMMODITIES

INDEX OF CLASSES, TYPES AND GROUPS OF COMMODITIES

INDEX OF GROUP LETTER CODES
IN ALPHABETICAL ORDER

INDEX OF GROUP LETTER CODES (IN ALPHABETICAL ORDER)

Group Letter Code	Group	Class	Type	Group No.	page No.
AB	By-products, used for animal feeding purposes, derived from fruit and vegetable processing	D	13	071	350
AF	Straw, fodder and forage of cereal grains and grasses (including buckwheat fodder) (forage)	C	11	051	328
AL	Legume animal feeds	C	11	050	324
AM	Miscellaneous Fodder and Forage crops (fodder)	C	11	052	331
AR	Frogs, lizards, snakes and turtles	B	09	048	318
AS	Straw, fodder and forage of cereal grains and grasses (including buckwheat fodder) (straws and fodder dry)	C	11	051	329
AV	Miscellaneous Fodder and Forage crops (forage)	C	11	052	331
CF	Cereal grain milling fractions	D	13	065	340
CM	Milled cereal products (early milling stages)	D	12	058	338
CP	Manufactured multi-ingredient cereal products	D	15	078	351
DF	Dried fruits	D	12	055	333
DH	Dried herbs	D	12	057	335
DM	Miscellaneous derived edible products of plant origin	D	13	069	348
DT	Teas	D	13	066	342
DV	Dried vegetables	D	12	056	335
FA	Animal fats, processed	E	17	085	358
FB	Berries and other small fruits	A	01	004	206
FC	Citrus fruits	A	01	001	196
FI	Assorted tropical and sub-tropical fruit - inedible peel	A	01	006	213
FM	Milk fats	E	17	086	360
FP	Pome fruits	A	01	002	201
FS	Stone fruits	A	01	003	203
FT	Assorted tropical and sub-tropical fruit - edible peel	A	01	005	209
GC	Cereal grains	A	03	020	256
GS	Grasses for sugar or syrup production	A	03	021	261
HH	Herbs	A	05	027	269
HS	Spices	A	05	028	274
IM	Molluscs (including Cephalopods) and other invertebrate animals	B	10	049	320
JF	Fruit juices	D	13	070	349
LD	Derived milk products	E	17	087	361
LI	Manufactured milk products (single-ingredient)	E	18	090	362

INDEX OF GROUP LETTER CODES (IN ALPHABETICAL ORDER)

Group Letter Code	Group	Class	Type	Group No.	page No.
LM	Manufactured milk produts (multi-ingredient)	E	19	092	363
LS	Secondary milk products	E	16	082	356
MD	Dried meat and fish products	E	16	080	353
MF	Mammalian fats	B	06	031	282
ML	Milks	B	06	033	285
MM	Meat (from mammals other than marine mammals)	B	06	030	278
MO	Edible offal (mammalian)	B	06	032	283
OC	Vegetable oils, crude	D	13	067	344
OR	Vegetable oils, edible (or refined)	D	13	068	346
PE	Eggs	B	07	039	290
PF	Poultry fats	B	07	037	288
PO	Poultry, Edible offal of	B	07	038	289
SB	Seed for beverages and sweets	A	04	024	268
SC	Crustaceans, processed	E	17	084	356
SM	Miscellaneous secondary food Commodities of plant origin	D	12	059	339
SO	Oilseed	A	04	023	265
TN	Tree nuts	A	04	022	262
VA	Bulb vegetables	A	02	009	218
VB	Brassica (cole or cabbage) vegetables, Head cabbage, Flowerhead brassicas	A	02	010	221
VC	Fruiting vegetables, Cucurbits	A	02	011	223
VD	Pulses	A	02	015	244
VL	Leafy vegetables (including brassica leafy vegetables)	A	02	013	231
VO	Fruiting vegetables, other than Cucurbits	A	02	012	228
VP	Legume vegetables	A	02	014	239
VR	Root and tuber vegetables	A	02	016	248
VS	Stalk and stem vegetables	A	02	017	254
WC	Crustaceans	B	08	045	314
WD	Diadromous fish	B	08	041	295
WF	Freshwater fish	B	08	040	291
WL	Fish roe (including milt = soft roe) and edible offal of fish: offal	B	08	043	309
WM	Marine mammals	B	08	044	311
WR	Fish roe (including milt = soft roe) and edible offal of fish: roe	B	08	043	309

INDEX OF FOOD AND ANIMAL FEED COMMODITIES
(IN ALPHABETICAL ORDER)

INDEX OF FOOD AND ANIMAL FEED COMMODITIES
(In Alphabetical order)

Class	Group			
A	005	FT	4095	Acerola, see Barbados cherry
A	020	GC	4597	Acha, see Hungry Rice
A	006	FI	5298	Achiote, see Annatto
A	016	VR	4527	Achira, see Canna, edible
A	020	GC	4599	Adlay, see Job's Tears
A	015	VD	560	Adzuki bean (dry)
A	020	GC	4601	African millet, see Millet, Finger
A	006	FI	325	Akee apple
B	042	WS	4937	Albacore, see Subgroup Tuna and Bonito
C	050	AL	1020	Alfalfa fodder
C	050	AL	1021	Alfalfa forage (green)
A	012	VO	4265	Alkekengi, see Ground cherries
A	028	HS	4769	Allspice fruit, see Pimento
A	022	TN	660	Almonds
A	016	VR	570	Alocasia
A	013	VL	460	Amaranth
A	005	FT	285	Ambarella
A	013	VL	4313	Amsoi, see Indian Mustard
B	040	WF	4837	Amur pike, see Pike
B	042	WS	920	Anchovies
A	028	HS	720	Angelica seed
A	027	HH	720	Angelica, including Garden Angelica
D	057	DH	720	Angelica, including Garden Angelica, dry
A	028	HS	4771	Angelica, root, stem and leaves, see Group 027: Herbs, Angelica, including Gar
A	014	VP	4393	Angola pea (immature seed), see Pigeon pea
A	015	VD	4465	Angola pea, see Pigeon pea
A	028	HS	771	Anise seed
A	028	HS	4773	Aniseed, see Anise seed
A	005	FT	4097	Aonla, see Otaheite gooseberry
A	002	FP	226	Apple
D	070	JF	226	Apple juice
D	071	AB	226	Apple pomace, dry
D	055	DF	226	Apples, dried
A	003	FS	240	Apricot
D	055	DF	240	Apricots, dried
A	005	FT	286	Arbutus berry
B	041	WD	4867	Arctic char, see Trout
A	016	VR	571	Arracacha
A	016	VR	572	Arrowhead
A	016	VR	573	Arrowroot
A	013	VL	4315	Arrugula, see Rucola
A	017	VS	620	Artichoke, Globe
A	017	VS	621	Asparagus
A	014	VP	4395	Asparagus bean (pods), see Yard-long bean
A	014	VP	4397	Asparagus pea (pods), see Goa bean
A	005	FT	26	Assorted tropical and sub-tropical fruits - edible peel
A	006	FI	30	Assorted tropical and sub-tropical fruits-inedible peel
B	041	WD	4877	Atlantic salmon, see Salmon, Atlantic
A	012	VO	4267	Aubergine, see Egg plant
A	006	FI	326	Avocado

INDEX OF FOOD AND ANIMAL FEED COMMODITIES
(In Alphabetical order)

Class	Group			
A	027	HH	721	Balm leaves
D	057	DH	721	Balm leaves, dry
A	011	VC	420	Balsam apple
A	011	VC	421	Balsam pear
A	013	VL	421	Balsam pear leaves
A	015	VD	520	Bambara groundnut (dry seed)
A	014	VP	520	Bambara groundnut (immature seeds)
A	017	VS	622	Bamboo shoots
A	006	FI	327	Banana
A	006	FI	328	Banana, Dwarf
A	005	FT	287	Barbados cherry
B	040	WF	855	Barbs
A	020	GC	640	Barley
C	051	AS	640	Barley straw and fodder, dry
B	042	WS	921	Barracudas
B	041	WD	898	Barramundi
A	027	HH	722	Basil
D	057	DH	722	Basil, dry
A	027	HH	723	Bay leaves
D	057	DH	7	Bay leaves, dry
C	050	AL	61	Bean fodder
C	050	AL	1030	Bean forage (green)
C	050	AL	1022	Bean, Velvet
A	015	VD	71	Beans (dry)
A	014	VP	61	Beans, except broad bean and soya bean
A	014	VP	62	Beans, shelled
A	004	FB	260	Bearberry
E	080	MD	5297	Beche-de-mer, dried, see Sea-cucumbers, dried
B	049	IM	5163	Beche-de-mer, see Sea-cucumbers
A	022	TN	661	Beech nuts
A	013	VL	4317	Beet leaves, see Chard
A	016	VR	574	Beetroot
A	012	VO	4269	Bell pepper, see Peppers, Sweet
A	023	SO	690	Ben Moringa seed
C	051	AS	5241	Bermuda grass
A	004	FB	18	Berries and other small fruits
A	013	VL	461	Betel leaves
A	001	FC	4000	Bigarade, see Orange, Sour
B	042	WS	4939	Bigeye tuna, see Tuna, Bigeye
A	004	FB	261	Bilberry
A	004	FB	262	Bilberry, Bog
A	004	FB	263	Bilberry, Red
A	005	FT	288	Bilimbi
A	013	VL	4319	Bitter cucumber leaves, see Balsam pear leaves
A	011	VC	4193	Bitter cucumber, see Balsam pear
A	011	VC	4195	Bitter gourd, see Balsam pear
A	011	VC	4197	Bitter melon, see Balsam pear
B	040	WF	856	Black bass
D	070	JF	1140	Black currant juice
A	015	VD	521	Black gram

INDEX OF FOOD AND ANIMAL FEED COMMODITIES
(In Alphabetical order)

Class	Group				
A	014	VP	521	Black gram (green pods)
A	016	VR	4529	Black salsify, see Scorzonera
A	015	VD	4467	Black-eyed pea, see Cowpea
A	004	FB	264	Blackberries
B	042	WS	4940	Blackfin tuna, see Tuna, Blackfin
A	013	VL	4321	Blackjack
A	013	VL	4323	Bledo, see Amaranth
A	001	FC	4001	Blood orange, see Orange, Sweet
A	004	FB	20	Blueberries
A	004	FB	4073	Blueberry, Highbush, see Blueberries
A	004	FB	4075	Blueberry, Lowbush, see Blueberries
A	004	FB	4077	Blueberry, Rabbiteye, see Blueberries
B	042	WS	922	Bluefish
B	043	WR	922	Bluefish roe (m)
B	040	WF	857	Bluegill sunfish (or Bluegill bream)
C	051	AS	5243	Bluegrass
B	042	WS	923	Bogue
A	014	VP	4399	Bonavist bean (young pods and immature seeds),
A	015	VD	4469	Bonavist bean, see Hyacinth bean
B	042	WS	924	Bonito
B	042	WS	4941	Bonito, Atlantic, see Bonito
B	042	WS	4943	Bonito, Eastern Pacific, see Bonito
A	027	HH	724	Borage
D	057	DH	724	Borage, dry
A	013	VL	4325	Borecole, see Kale, curly
A	011	VC	422	Bottle gourd
A	013	VL	462	Box thorn
A	004	FB	4079	Boysenberry, see Dewberries
D	058	CM	81	Bran, unprocessed of cereal grain
A	010	VB	40	Brassica (cole or cabbage) vegetables, Head cabbages, Flowerhead brassicas
A	013	VL	54	Brassica leafy vegetables
A	022	TN	662	Brazil nut
A	005	FT	4099	Brazilian cherry, see Grumichana
D	078	CP	179	Bread and other cooked cereal products
A	006	FI	329	Breadfruit
B	040	WF	858	Bream
B	042	WS	956	Bream, Silver
B	042	WS	4945	Brill, see Turbot
A	015	VD	523	Broad bean (dry)
A	014	VP	522	Broad bean (green pods and immature seeds)
A	014	VP	523	Broad bean, shelled (succulent) (= immature seeds)
A	010	VB	400	Broccoli
A	013	VL	4327	Broccoli raab
A	010	VB	401	Broccoli, Chinese
A	010	VB	4173	Broccoli, Sprouting, see Broccoli
C	051	AS	5245	Brome grass
B	041	WD	4869	Brook trout, see Trout
B	041	WD	4871	Brown trout, see Trout
B	040	WF	4839	Brown trout, see Trout, Brown
A	020	GC	4603	Brown-corn millet, see Millet, Common

INDEX OF FOOD AND ANIMAL FEED COMMODITIES
(In Alphabetical order)

Class	Group			
A	010	VB	402	Brussels sprouts
A	020	GC	641	Buckwheat
C	051	AS	641	Buckwheat fodder
B	031	MF	810	Buffalo fat
B	030	MM	810	Buffalo meat
B	033	ML	810	Buffalo milk
E	086	FM	810	Buffalo milk fat
E	085	FA	810	Buffalo tallow
B	030	MM	4789	Buffalo, African, meat, see Buffalo meat
B	030	MM	4791	Buffalo, American, meat, see Buffalo meat
B	030	MM	4793	Buffalo, Cape, meat, see Buffalo, African, meat
B	032	MO	810	Buffalo, Edible offal of
B	030	MM	4795	Buffalo, Water, meat, see Buffalo meat
A	009	VA	35	Bulb vegetables
A	009	VA	36	Bulb vegetables, except Fennel, Bulb
B	048	AR	5145	BullFrog, Indian, see Frogs
A	003	FS	241	Bullace
B	048	AR	5143	Bullfrog, see Frogs
A	020	GC	4607	Bulrush millet, see Millet, Bulrush
A	016	VR	575	Burdock, greater or edible
A	027	HH	725	Burnet, Great
A	027	HH	4731	Burnet, Salad, see Burnet, Great
A	027	HH	720	Burning bush
D	057	DH	728	Burning bush, dry
A	022	TN	4681	Bush nut, see Macadamia nut
A	014	VP	4401	Butter bean (immature pods), see Lima bean
A	015	VD	4470	Butter bean, see Lima bean
A	022	TN	663	Butter nut
B	042	WS	925	Butterfish
A	010	VB	4177	Cabbage, Green, see Cabbage, Savoy
A	010	VB	4181	Cabbage, Oxhead, see Cabbages Head
A	010	VB	4183	Cabbage, Pointed, see Cabbage, Oxhead
A	010	VB	4179	Cabbage, Red, see Cabbages, Head
A	010	VB	403	Cabbage, Savoy, see also Cabbages, Head
A	010	VB	4185	Cabbage, White, see Cabbages, Head
A	010	VB	4187	Cabbage, Yellow, see Cabbage, Savoy
A	010	VB	4175	Cabbage, see Cabbages, Head
A	010	VB	41	Cabbages, Head
A	024	SB	715	Cacao beans
A	014	VP	4401	Cajan pea (young green seeds), see Pigeon pea
A	015	VD	4471	Cajan pea, see Pigeon pea
A	001	FC	201	Calamondin, see also Subgroup 3 Mandarins
A	028	HS	772	Calamus, root
B	030	MM	4797	Calf meat, see Cattle meat
B	031	MF	811	Camel fat
B	030	MM	811	Camel meat
B	033	ML	811	Camel milk
E	086	FM	811	Camel milk fat
E	085	FA	811	Camel tallow
B	030	MM	4799	Camel, Bactrian, meat, see Camel meat

INDEX OF FOOD AND ANIMAL FEED COMMODITIES
(In Alphabetical order)

Class	Group			
B	032	MO	811	Camel, Edible offal of
D	066	DT	1110	Camomile or Chamomile
D	066	DT	5277	Camomile, German or Scented, see Camomile
D	066	DT	5279	Camomile, Roman or Noble, see Camomile
A	006	FI	330	Canistel
A	016	VR	576	Canna, edible
A	011	VC	4199	Cantaloupe, see Melons
A	012	VO	4271	Cape gooseberry, see Ground cherries
B	042	WS	926	Capelin
A	028	HS	773	Caper buds
A	005	FT	289	Carambola
A	005	FT	290	Caranda
A	028	HS	774	Caraway seed
A	028	HS	775	Cardamom seed
A	017	VS	623	Cardoon
A	005	FT	291	Carob
A	009	VA	4153	Carosella, see Fennel, Italian
B	040	WF	4843	Carp, Chinese, see Carp, Grass
B	040	WF	4841	Carp, Common, see Carps
B	040	WF	4845	Carp, Grass, see Carps
B	040	WF	860	Carp, Indian
B	040	WF	859	Carps
A	016	VR	577	Carrot
A	011	VC	4201	Casaba or Casaba melon, see Subgroup Melons, except Watermelon
A	005	FT	292	Cashew apple
A	022	TN	295	Cashew nut
A	016	VR	463	Cassava
A	013	VL	463	Cassava leaves
A	016	VR	4531	Cassava, Bitter, see Cassava
A	016	VR	4533	Cassava, Sweet, see Cassava
A	028	HS	4775	Cassia bark, see Cinnamon bark (including Cinnamon, Chinese bark)
A	028	HS	776	Cassia buds
D	070	JF	5293	Cassis, see Black currant juice
A	020	GC	4609	Cat-tail millet, see Millet, Bulrush
B	042	WS	4947	Catfish, Sea, see Wolffish
B	040	WF	861	Catfishes (freshwater)
A	014	VP	4401	Catjang cowpea (immature pods and green seeds), see Cowpea
A	027	HH	726	Catmint
D	057	DH	726	Catmint, dry
A	027	HH	4733	Catnip, see Catmint
B	031	MF	812	Cattle fat
B	030	MM	812	Cattle meat
E	080	MD	812	Cattle meat, dried (including dried and smoked)
B	033	ML	812	Cattle milk
E	086	FM	812	Cattle milk fat
E	085	FA	812	Cattle tallow (including processed suet)
B	032	MO	812	Cattle, Edible offal of
B	032	MO	1280	Cattle, Kidney
B	032	MO	1281	Cattle, liver
A	010	VB	4189	Cauliflower, Green, see Cauliflower

INDEX OF FOOD AND ANIMAL FEED COMMODITIES
(In Alphabetical order)

Class	Group			
A	010	VB	404	Cauliflower, see also Flowerhead brassicas
A	020	GC	642	CaÊihua
A	016	VR	578	Celeriac
A	017	VS	624	Celery
A	013	VL	4329	Celery cabbage, see Chinese cabbage
A	027	HH	624	Celery leaves
D	057	DH	624	Celery leaves, dry
A	017	VS	4595	Celery leaves, see Group 027: Herbs
A	013	VL	4331	Celery mustard, see Pak-choi
A	028	HS	624	Celery seed
A	017	VS	625	Celtuce
B	049	IM	152	Cephalopods
D	065	CF	81	Cereal brans, processed
A	020	GC	80	Cereal grains
A	020	GC	81	Cereal grains, except Buckwheat, CaÊihua and Quinoa
B	040	WF	4847	Channel catfish, see Catfishes (freshwater)
B	041	WD	4873	Char, see Lake trout
A	013	VL	464	Chard
A	011	VC	423	Chayote
A	016	VR	423	Chayote root
A	006	FI	331	Cherimoya
A	003	FS	13	Cherries
A	012	VO	4273	Cherry pepper, see Peppers, Chili
A	003	FS	242	Cherry plum
B	041	WD	4875	Cherry salmon, See Subgroup Salmon, Pacific
A	012	VO	4275	Cherry tomato, see Ground cherries
A	003	FS	243	Cherry, Sour
A	003	FS	244	Cherry, Sweet
A	013	VL	465	Chervil
A	016	VR	579	Chervil, Turnip-rooted
A	027	HH	4735	Chervil, see Group 013: Leafy vegetables
A	022	TN	664	Chestnuts
A	015	VD	524	Chick-pea (dry)
A	014	VP	524	Chick-pea (green pods)
C	050	AL	524	Chick-pea fodder
A	003	FS	4053	Chickasaw plum, see Plum, Chickasaw
A	020	GC	4611	Chicken corn, see Sorghum
B	039	PE	840	Chicken eggs
B	037	PF	840	Chicken fat
E	085	FA	840	Chicken fat, processed
B	036	PM	840	Chicken meat
B	038	PO	840	Chicken, Edible offal of
C	050	AL	5217	Chickling vetch, see Vetch, Chickling
A	013	VL	469	Chicory leaves (green and red cultivars
A	016	VR	469	Chicory, roots
A	012	VO	4277	Chili peppers, see Peppers, Chili
A	013	VL	467	Chinese cabbage, (type Pe-tsai)
A	006	FI	4127	Chinese gooseberry, see Kiwifruit
A	012	VO	4279	Chinese lantern plant, see Ground cherries
A	005	FT	293	Chinese olive, Black, White

INDEX OF FOOD AND ANIMAL FEED COMMODITIES
(In Alphabetical order)

Class	Group			
A	006	FI	4128	Chinese persimmon, see subgroup 5 Persimmon, Japanese
A	016	VR	4535	Chinese radish, see Radish, Japanese
B	041	WD	4889	Chinook salmon, see Subgroup Salmon, Pacific
A	001	FC	4002	Chinotto, see Orange, Sour
A	022	TN	4683	Chinquapin, see Chestnuts
A	001	FC	4003	Chironja, see Subgroup Oranges, Sweet, Sour (including Orange-like hybrids)
A	027	HH	727	Chives
A	027	HH	4737	Chives, Chinese, see Chives
A	009	VA	4157	Chives, Chinese, see Group 027: Herbs
A	009	VA	4155	Chives, see Group 027: Herbs
A	013	VL	468	Choisum
A	011	VC	4203	Christophine, see Chayote
A	016	VR	4537	Christophine, see Chayote root
A	016	VR	580	Chufa, see Tiger nut
B	041	WD	4891	Chum salmon, see Subgroup Salmon, Pacific
A	028	HS	777	Cinnamon bark (including Cinnamon, Chinese bark)
A	011	VC	4205	Citron melon, see Watermelon
A	001	FC	202	Citron, see also Subgroup 2 Lemons and Limes
A	001	FC	1	Citrus fruits
D	070	JF	1	Citrus juice
D	069	DM	1	Citrus molasses
D	071	AB	1	Citrus pulp, dry
B	049	IM	1000	Clams
A	027	HH	4739	Clary, see Sage (and related Salvia species)
A	001	FC	4005	Clementine, see Mandarins
A	001	FC	4006	Cleopatra mandarin, see Subgroup 3 Mandarins
A	004	FB	277	Cloudberry
C	050	AL	1023	Clover
C	050	AL	1031	Clover hay or fodder
A	028	HS	778	Cloves, buds
A	014	VP	525	Cluster bean (young pods)
A	012	VO	4281	Cluster pepper, see Peppers, Chili
B	042	WS	4949	Coalfish, see Pollack
B	049	IM	5165	Cockle, Common, see Cockles
B	049	IM	1001	Cockles
A	005	FT	294	Coco plum
D	069	DM	1215	Cocoa butter
D	069	DM	1216	Cocoa mass
D	069	DM	715	Cocoa powder
A	022	TN	665	Coconut
D	067	OC	665	Coconut oil, crude
D	068	OR	665	Coconut oil, refined
A	023	SO	4701	Coconut, see Group 022: Tree nuts
A	016	VR	4539	Cocoyam, see Tannia and Taro
B	042	WS	927	Cod
B	042	WS	126	Cod and Cod-like fishes
B	043	WL	927	Cod liver (m)
B	043	WR	927	Cod roe (m)
B	042	WS	4951	Cod, Atlantic, see Cod
B	042	WS	4953	Cod, Greenland, see Cod

INDEX OF FOOD AND ANIMAL FEED COMMODITIES
(In Alphabetical order)

Class	Group			
B	040	WF	869	Cod, Murray
B	042	WS	4955	Cod, Pacific, see Cod
E	080	MD	927	Cod, dried
A	024	SB	716	Coffee beans
B	041	WD	4893	Coho salmon, see Subgroup Salmon, Pacific
A	024	SB	717	Cola nuts
A	013	VL	4332	Collard, see Kale
A	023	SO	4705	Colza, Indian, see Mustard seed, Field
A	023	SO	4703	Colza, see Rape seed
A	015	VD	526	Common bean (dry)
A	014	VP	526	Common bean (pods and/or immature seeds)
A	012	VO	4283	Cone pepper, see peppers, Chili
B	042	WS	928	Conger or Conger eel
B	042	WS	4957	Conger, European, see Conger
A	028	HS	779	Coriander, seed
D	078	CP	5295	Corn bread, see Maize bread
D	065	CF	5273	Corn flour, see Maize flour
C	051	AS	5247	Corn fodder, see Maize fodder
C	051	AF	5249	Corn forage, see Maize forage
D	065	CF	5275	Corn meal, see Maize meal
D	067	OC	5289	Corn oil, crude, see Maize oil, crude
D	068	OR	5291	Corn oil, edible, see Maize oil, edible
A	013	VL	470	Corn salad
A	020	GC	4613	Corn, see Maize
A	020	GC	4617	Corn, whole kernel (Codex Stan. 132-1981, see Fruiting vegetables (other tha
A	020	GC	4615	Corn-on-the-cob (Codex Stan. 133-1981), see Fruiting vegetables (other than
A	012	VO	4285	Corn-on-the-cob, see Sweet corn (corn-on-the-cob)
A	013	VL	510	Cos lettuce
A	027	HH	4741	Costmary, see Tansy (and related species)
C	052	AM	691	Cotton fodder, dry
A	023	SO	691	Cotton seed
D	067	OC	691	Cotton seed oil, crude
D	068	OR	691	Cotton seed oil, edible
A	011	VC	4207	Courgette, see Squash, Summer
C	052	AV	1050	Cow cabbage
A	004	FB	4081	Cowberry, see Bilberry, Red
A	015	VD	527	Cowpea (dry)
A	014	VP	527	Cowpea (immature pods)
A	013	VL	4333	Cowslip, see Marsh marigold
A	002	FP	227	Crab-apple
E	084	SC	146	Crabmeat, cooked
B	045	WC	146	Crabs
A	004	FB	265	Cranberry
A	013	VL	472	Cress, Garden
D	057	DH	5269	Cretan Dittany, dry, see Burning bush, dry
A	027	HH	4743	Cretan Dittany, see Burning bush
A	013	VL	4335	Crisphead lettuce, see Lettuce, Head
B	045	WC	143	Crustaceans
E	084	SC	143	Crustaceans, cooked
A	011	VC	424	Cucumber

INDEX OF FOOD AND ANIMAL FEED COMMODITIES
(In Alphabetical order)

Class	Group			
A	011	VC	4209	Cucuzzi, see Bottle Gourd
A	028	HS	780	Cumin seed
A	013	VL	4337	Curly kale, see Kale, curly
A	004	FB	278	Currant, Black, see also Currants, Black, Red, White
A	004	FB	279	Currant, Red, White, see also Currants, Black, Red, White
D	055	DF	5257	Currants
A	004	FB	21	Currants, Black, Red, White
A	027	HH	729	Curry leaves
A	011	VC	4211	Cushaws, see Pumpkins
A	006	FI	332	Custard apple
B	041	WD	4895	Cutthroat trout, see Trout
A	013	VL	4339	Cutting lettuce, see Lettuce, leaf
B	049	IM	5167	Cuttlefish, Common, see Cuttlefishes
B	049	IM	1002	Cuttlefishes
B	042	WS	929	Dab or Common dab
E	080	MD	929	Dab or Common dab, dried
A	016	VR	4543	Daikon, see Radish, Japanese
A	003	FS	4055	Damsons (Damson plums) see Plum, Damson
A	001	FC	4007	Dancy or Dancy mandarin, see Subgroup 3 Mandarins
A	013	VL	474	Dandelion
A	020	GC	4619	Dari seed, see Sorghum
C	051	AS	5251	Darnel
A	016	VR	4541	Dasheen, see Taro
A	005	FT	295	Date
D	055	DF	295	Dates, dried or dried and candied
B	030	MM	813	Deer meat
B	030	MM	4803	Deer, Fallow, meat, see Deer meat
B	030	MM	4805	Deer, Red, meat, see Deer meat
A	005	FT	296	Desert date
A	023	SO	4707	Desert date, see Group 5: Assorted tropical and sub-tropical fruits - edible peel
A	004	FB	266	Dewberries (including Boysenberry and Loganberry)
B	041	WD	120	Diadromous fish
E	080	MD	120	Diadromous fish, dried
A	027	HH	730	Dill
A	028	HS	730	Dill seed
A	013	VL	475	Dock
B	044	WM	5045	Dolphin, Bottlenose, see Dolphins
B	044	WM	5047	Dolphin, Humpback, see Dolphins
B	044	WM	5049	Dolphin, Spinner, see Dolphins
B	042	WS	930	Dolphinfish
B	043	WR	930	Dolphinfish roe (m)
B	044	WM	970	Dolphins
B	042	WS	4959	Dorado, see Dolphinfish
A	006	FI	333	Doum or Dum palm
D	056	DV	168	Dried Vegetables
E	080	MD	180	Dried fish
D	055	DF	167	Dried fruits
D	055	DF	269	Dried grapes (= Currants, Raisins and Sultanas)
D	057	DH	170	Dried herbs
D	055	DF	5259	Dried vine fruits, see Dried grapes

INDEX OF FOOD AND ANIMAL FEED COMMODITIES
(In Alphabetical order)

Class	Group			
B	030	MM	4807	Dromedary meat, see Camel meat
B	042	WS	931	Drums
A	023	SO	4709	Drumstick tree seed, see Ben Moringa seed
B	039	PE	841	Duck eggs
B	037	PF	841	Duck fat
E	085	FA	841	Duck fat, processed
B	036	PM	841	Duck meat
B	038	PO	841	Duck, Edible offal of
A	006	FI	334	Durian
A	020	GC	4621	Durra, see Sorghum
A	020	GC	4623	Durum wheat, see Wheat
A	015	VD	4473	Dwarf bean (dry), see Common bean (dry)
A	014	VP	4403	Dwarf bean (immature pods and/or seeds), see Common bean
A	016	VR	4545	Eddoe, see Taro
B	032	MO	105	Edible offal (Mammalian)
B	032	MO	96	Edible offal of cattle, goats, horses, pigs and sheep
B	032	MO	97	Edible offal of cattle, pigs and sheep
A	014	VP	4405	Edible-podded pea, see Podded pea
B	041	WD	4897	Eel, American, see Eels
B	041	WD	4899	Eel, Australian, see Eels
B	041	WD	4901	Eel, European, see Eels
B	041	WD	4903	Eel, Japanese, see Eels
B	041	WD	890	Eels
A	006	FI	4129	Egg fruit, see Canistel
A	012	VO	440	Egg plant
B	039	PE	112	Eggs
A	004	FB	267	Elderberries
A	028	HS	781	Elecampane, root
A	006	FI	371	Elephant apple
B	030	MM	824	Elk meat
A	020	GC	4625	Emmer, see Wheat
A	013	VL	476	Endive
A	013	VL	4341	Endive, broad or plain leaved, see Endive
A	013	VL	4343	Endive, curled, see Endive
A	027	HH	4745	Estragon, see Tarragon
B	042	WS	4961	European sardine, see Subgroup Sardines and Sardine fishes
B	044	WM	142	Fat of Dolphins, Seals and Whales, unprocessed
A	015	VD	4475	Fava bean (dry), see Broad bean (dry)
A	014	VP	4407	Fava bean (green pods and immature beans), see Broad bean
A	006	FI	335	Feijoa
A	027	HH	731	Fennel
A	009	VA	380	Fennel, Bulb
A	027	HH	4747	Fennel, Bulb, see Group 01: Bulb vegetables, No. VA 0380
A	013	VL	4347	Fennel, Bulb, see Group 9 Bulb vegetables
A	009	VA	4159	Fennel, Italian, see Fennel, bulb
A	009	VA	4161	Fennel, Roman, see Fennel, bulb
A	009	VA	4163	Fennel, Sweet, see Fennel, Roman
D	057	DH	731	Fennel, dry
A	013	VL	4345	Fennel, see Group 027 Herbs
A	028	HS	731	Fennel, seed

INDEX OF FOOD AND ANIMAL FEED COMMODITIES
(In Alphabetical order)

Class	Group			
A	028	HS	782	Fenugreek, seed
C	051	AS	5253	Fescue
A	020	GC	4627	Feterita, see Sorghum
A	015	VD	4477	Field bean (dry), see Common bean (dry)
A	014	VP	4409	Field bean (green pods), see Common bean
A	015	VD	561	Field pea (dry)
A	005	FT	297	Fig
D	055	DF	297	Figs, dried or dried and candied
A	022	TN	4685	Filberts, see Hazelnuts
A	020	GC	4629	Finger millet, see Millet, Finger
B	043	WR	140	Fish roe
A	015	VD	4479	Flageolet (dry), see Common bean (dry)
A	014	VP	4411	Flageolet (fresh beans), see Common bean
B	042	WS	127	Flat-fishes
E	080	MD	127	Flat-fishes, dried
A	023	SO	4711	Flax-seed, see Linseed
B	043	WR	932	Flounder roe (m)
B	042	WS	932	Flounders
A	010	VB	42	Flowerhead brassicas (includes Broccoli: Broccoli, Chinese and Cauliflower)
C	052	AM	1051	Fodder beet
C	052	AV	1051	Fodder beet leaves or tops
A	020	GC	4631	Fonio, see Hungry Rice
A	014	VP	4413	Four-angled bean (immature pods), see Goa bean
A	020	GC	4633	Foxtail millet, see Millet, Foxtail
A	014	VP	4415	French bean (immature pods and seeds), see Common bean
A	015	VD	4481	French bean, see Group 06: Legume vegetables
B	045	WC	976	Freshwater crayfishes
E	084	SC	976	Freshwater crayfishes, cooked
B	045	WC	144	Freshwater crustaceans
E	084	SC	144	Freshwater crustaceans, cooked
B	040	WF	115	Freshwater fish
B	045	WC	977	Freshwater shrimps or prawns
E	084	SC	977	Freshwater shrimps or prawns, cooked
B	048	AR	5147	Frog, Agile, see Frogs
B	048	AR	5149	Frog, Common, see Frogs
B	048	AR	5151	Frog, Edible, see Frogs
B	048	AR	5153	Frog, Marsh, see Frogs
B	048	AR	5155	Frog, Pool, see Frogs
B	048	AR	990	Frogs
B	048	AR	148	Frogs, lizards, snakes and turtles
D	070	JF	175	Fruit juices
A	011	VC	45	Fruiting vegetables, Cucurbits
A	012	VO	50	Fruiting vegetables, other than Cucurbits
A	020	GC	4635	Fundi, see Hungry Rice
A	012	VO	449	Fungi, Edible (not including mushrooms)
A	012	VO	4287	Fungus "Chanterelle", see Fungi, Edible
A	016	VR	581	Galangal, Greater
A	016	VR	582	Galangal, Lesser
A	028	HS	783	Galangal, rhizomes
A	014	VP	4417	Garbanzos, see Chick-pea

INDEX OF FOOD AND ANIMAL FEED COMMODITIES
(In Alphabetical order)

Class	Group			
A	013	VL	4349	Garden cress, see Cress, Garden
A	014	VP	528	Garden pea (young pods) (= succulent, immature seeds)
A	015	VD	4485	Garden pea, see Group 06: Legume vegetables
A	014	VP	529	Garden pea, shelled (succulent seeds)
B	042	WS	933	Garfish
A	009	VA	381	Garlic
A	009	VA	382	Garlic, Great-headed
A	006	FI	4131	Genip, see Marmaladebox
A	015	VD	4483	Geocarpa groundnut or Geocarpa bean, see Kersting's groundnut
B	041	WD	4907	German trout, see Trout
A	011	VC	425	Gherkin
A	011	VC	426	Gherkin, West Indian
B	041	WD	4905	Giant sea perch, see Barramundi
B	049	IM	5169	Giant snail, see Snails, Edible (Africa, Asia)
A	028	HS	784	Ginger, root
A	016	VR	4547	Globe artichoke, see Group 09: Artichoke Globe, Stalk and stem vegetables
A	014	VP	530	Goa bean (immature pods)
A	016	VR	530	Goa bean root
A	015	VD	4487	Goa bean, see Group 06: Legume vegetables
B	031	MF	814	Goat fat
B	030	MM	814	Goat meat
B	033	ML	814	Goat milk
E	086	FM	814	Goat milk fat
E	085	FA	814	Goat tallow
B	032	MO	14	Goat, Edible offal of
B	040	WF	862	Gobies, Freshwater
A	012	VO	4289	Golden berry, see Ground cherries
E	085	FA	842	Goose Fat, processed
B	039	PE	842	Goose eggs
B	037	PF	842	Goose fat
B	036	PM	842	Goose meat
B	038	PO	842	Goose, Edible offal of
B	038	PO	849	Goose, liver
A	004	FB	268	Gooseberry
A	013	VL	477	Goosefoot
B	040	WF	863	Gourami (Asia)
A	013	VL	4351	Gow Kee, see Box thorn
A	028	HS	785	Grains of paradise
A	015	VD	4489	Gram (dry), see Chick-pea (dry)
A	014	VP	4419	Gram (green pods), see Chick-pea
A	006	FI	4132	Granddilla, see Passion fruit
D	070	JF	269	Grape juice
A	013	VL	269	Grape leaves
D	071	AB	269	Grape pomace, dry
D	070	JF	203	Grapefruit juice
A	001	FC	203	Grapefruit, see also Subgroup 5 Shaddocks or Pomelos
A	004	FB	269	Grapes
C	050	AL	5219	Grass pea, see Vetch, Chickling
A	014	VP	4421	Green bean (green pods and immature seeds), see Common bean
A	015	VD	4491	Green beans, see Group 06: Legume vegetables

INDEX OF FOOD AND ANIMAL FEED COMMODITIES
(In Alphabetical order)

Class	Group			
A	015	VD	4493	Green gram (dry), see Mung bean (dry)
A	014	VP	4423	Green gram (green pods), see Mung bean
A	003	FS	4056	Greengages (Greengage plums), see Plum, Greengage
A	012	VO	441	Ground cherries
A	023	SO	4713	Groundnut, see Peanut
A	015	VD	4495	Groundnut, see Peanut, Group 023: Oilseed
A	005	FT	298	Grumichama
A	016	VR	4549	Gruya, see Canna, edible
A	006	FI	4134	Guanabana, see Soursop
A	014	VP	4425	Guar (young pods), see Cluster bean
A	006	FI	336	Guava
A	020	GC	4637	Guinea corn, see Sorghum
B	036	PM	843	Guinea-fowl meat
B	042	WS	934	Haddock
B	042	WS	935	Hakes
E	080	MD	935	Hakes, dried
B	042	WS	936	Halibut
B	042	WS	4963	Halibut, Atlantic, see Halibut
B	042	WS	4965	Halibut, Greenland, see Halibut
B	042	WS	4967	Halibut, Pacific, see Halibut
E	080	MD	936	Halibut, dried
B	031	MF	815	Hare fat
B	030	MM	815	Hare meat
A	014	VP	4427	Haricot bean (green pods, and/or immature seeds),
A	015	VD	4497	Haricot bean, see Common bean, Group 06: Legume vegetables
C	051	AS	162	Hay or fodder (dry) of grasses
A	022	TN	666	Hazelnuts
A	027	HH	726	Herbs
B	042	WS	937	Herring
B	043	WR	937	Herring roe (m)
B	042	WS	4969	Herring, Atlantic, see Herring
B	042	WS	4971	Herring, Pacific, see Herring
A	022	TN	667	Hickory nuts
A	020	GC	4639	Hog millet, see Millet, Common
A	005	FT	299	Hog plum
D	057	DH	1100	Hops, dry
A	027	HH	732	Horehound
D	057	DH	732	Horehound, dry
A	015	VD	4499	Horse bean (dry), see Broad bean (dry)
A	014	VP	4429	Horse bean (green pods and/or immature seeds),
B	031	MF	816	Horse fat
A	015	VD	562	Horse gram
B	042	WS	4973	Horse mackerel, see Jack Mackerel
B	030	MM	816	Horse meat
E	080	MD	816	Horse meat, dried (including dried and smoked)
E	085	FA	816	Horse tallow
B	032	MO	816	Horse, Edible offal of
B	032	MO	1292	Horse, kidney
B	032	MO	1293	Horse, liver
A	016	VR	583	Horseradish

INDEX OF FOOD AND ANIMAL FEED COMMODITIES
(In Alphabetical order)

Class	Group			
A	023	SO	4715	Horseradish tree seed, see Ben Moringa seed
A	028	HS	4779	Horseradish, see VR 0583, Group 016: Root and Tuber vegetables
A	004	FB	4083	Huckleberries
A	020	GC	643	Hungry rice
A	012	VO	4291	Husk tomato, see Ground cherries
A	015	VD	531	Hyacinth bean (dry)
A	014	VP	531	Hyacinth bean (young pods, immature seeds)
A	027	HH	733	Hyssop
D	057	DH	733	Hyssop, dry
A	005	FT	4101	Icaco plum, see Coco plum
A	006	FI	337	Ilama
A	006	FI	4133	Indian Fig, see Prickly pear
B	042	WS	4975	Indian mackerel, see Mackerel
A	013	VL	478	Indian mustard
A	006	FI	4136	Indian wood apple, see Elephant apple
A	005	FT	300	Jaboticaba
A	014	VP	532	Jack bean (young pods, immature seeds)
A	015	VD	4501	Jack bean, see Group 06: Legume vegetables
B	042	WS	938	Jack mackerel
A	006	FI	338	Jackfruit
A	013	VL	4353	Jamaican sorrel, see Roselle leaves
A	006	FI	339	Jambolan
A	016	VR	584	Japanese artichoke
A	009	VA	4165	Japanese bunching onion, see Welsh onion
A	013	VL	479	Japanese greens, various species, a.o.
A	022	TN	668	Japanese horse-chestnut
A	002	FP	4044	Japanese medlar, see Loquat
A	005	FT	4103	Java almond, see Group 024: Tree nuts
A	022	TN	4687	Java almonds, see Pili nuts
A	006	FI	340	Java apple
A	016	VR	585	Jerusalem artichoke
A	016	VR	4551	Jicama, see Yam bean
A	020	GC	644	Job's tears
A	005	FT	302	Jujube, Chinese
A	005	FT	301	Jujube, Indian
A	004	FB	270	Juneberries
A	028	HS	786	Juniper, berry
A	020	GC	4641	Kaffir corn, see Sorghum
A	010	VB	4191	Kailan, see Broccoli, Chinese
A	005	FT	4105	Kaki or Kaki fruit, see Persimmon, Japanese
A	013	VL	480	Kale (including among others: Collards, Curly kale, Scotch kale, thousand-he
A	013	VL	4355	Kale, curly, see Kale
B	030	MM	817	Kangaroo meat
A	013	VL	507	Kangkung
A	020	GC	4643	Kaoliang, see Sorghum
A	023	SO	692	Kapok
A	015	VD	563	Kersting's groundnut
B	041	WD	4909	Keta salmon, see Subgroup Salmon, Pacific
A	015	VD	4503	Kidney bean (dry), see Common bean (dry)
A	014	VP	4431	Kidney bean (pods and/or immature seeds), see Common bean

INDEX OF FOOD AND ANIMAL FEED COMMODITIES
(In Alphabetical order)

Class	Group			
B	032	MO	98	Kidney of cattle, goats, pigs and sheep
B	042	WS	939	King mackerel
A	001	FC	4008	King mandarin, see Subgroup 3 Mandarin
B	041	WD	4911	King salmon, see Subgroup Salmon, Pacific
A	006	FI	341	Kiwifruit
A	010	VB	405	Kohlrabi
A	024	SB	4727	Kola, see Cola nuts
A	013	VL	481	Komatsuma
C	050	AL	1024	Kudzu
C	050	AL	5221	Kudzu, Tropical, see Kudzu
A	005	FT	4107	Kumquat, Marumi, see Kumquats
A	005	FT	4109	Kumquat, Nagami, see Kumquats
A	005	FT	303	Kumquats
A	009	VA	383	Kurrat
A	015	VD	4505	Lablab (dry), see Hyacinth bean (dry)
A	014	VP	4433	Lablab (young pods; immature seeds), see Hyacinth bean
A	012	VO	4293	Lady's finger, see Okra
B	041	WD	4913	Lake trout, see Trout
B	030	MM	4809	Lamb meat, see Sheep meat
A	013	VL	4357	Lambs lettuce, see Corn salad
B	045	WC	5093	Langouste, see Spiny Lobster
E	085	FA	818	Lard (of pigs)
A	027	HH	734	Lavender
D	057	DH	734	Lavender, dry
A	013	VL	53	Leafy vegetables
A	009	VA	384	Leek
C	050	AL	157	Legume animal feeds
A	014	VP	60	Legume vegetables
D	066	DT	1111	Lemon verbena (dry leaves)
A	001	FC	204	Lemon, see also Subgroup 2 Lemons and Limes
A	001	FC	2	Lemons and Limes (including Citron)
A	015	VD	533	Lentil (dry)
A	014	VP	533	Lentil (young pods)
A	016	VR	4553	Leren, see Topee Tambu
C	050	AL	1025	Lespedeza
A	013	VL	482	Lettuce, Head
A	013	VL	483	Lettuce, Leaf
A	013	VL	4359	Lettuce, Red, see Lettuce, Head
A	028	HS	4781	Licorice, see Liquorice
A	015	VD	534	Lima bean (dry)
A	014	VP	534	Lima bean (young pods and/or immature beans)
D	066	DT	1112	Lime blossoms
A	001	FC	205	Lime, see also Subgroup 2 Lemons and Limes
B	042	WS	940	Ling
E	080	MD	940	Ling, dried
A	023	SO	693	Linseed
A	028	HS	787	Liquorice, roots
A	006	FI	343	Litchi
B	049	IM	5171	Little cuttle, see Cuttlefishes
B	032	MO	99	Liver of cattle, goats, pigs and sheep

INDEX OF FOOD AND ANIMAL FEED COMMODITIES
(In Alphabetical order)

Class	Group			
B	042	WS	4977	Liveroil shark, see Subgroup Sharks
B	048	AR	991	Lizards
B	030	MM	4811	Llama or Lama meat, see Camel meat
B	045	WC	5095	Lobster, American, see Lobsters
B	045	WC	5097	Lobster, European, see Lobsters
B	045	WC	5099	Lobster, Norway, see Lobsters
B	045	WC	978	Lobsters
E	084	SC	978	Lobsters (including Lobster meat), cooked
A	005	FT	4111	Locust tree, see Carob
A	004	FB	4085	Loganberry, see Dewberries
A	006	FI	342	Longan
B	042	WS	4979	Longtail tuna, see Tuna, Longtail
A	011	VC	427	Loofah, Angled
A	011	VC	428	Loofah, Smooth
A	002	FP	228	Loquat
A	027	HH	735	Lovage
D	057	DH	755	Lovage, dry
A	028	HS	735	Lovage, seed
A	006	FI	4135	Lulo, see Naranjilla
A	014	VP	545	Lupin
A	015	VD	545	Lupin (dry)
C	050	AL	545	Lupin, forage
A	022	TN	669	Macadamia nuts
A	028	HS	788	Mace
B	042	WS	941	Mackerel
B	042	WS	129	Mackerel and Jack Mackerel
B	042	WS	128	Mackerel and Mackerel-like Fishes
B	043	WR	941	Mackerel roe (m)
B	042	WS	4981	Mackerel, Atlantic, see Mackerel
B	042	WS	4983	Mackerel, Chub, see Mackerel
B	042	WS	4985	Mackerel, Indian, see Mackerel and Indian Mackerel
B	042	WS	4987	Mackerel, Short, see Mackerel
A	020	GC	645	Maize
D	078	CP	645	Maize bread
D	065	CF	1255	Maize flour
C	051	AS	645	Maize fodder
C	051	AF	645	Maize forage
D	065	CF	645	Maize meal
D	067	OC	645	Maize oil, crude
D	068	OR	645	Maize oil, edible
A	023	SO	4718	Maize, see Group 020: Cereal Grains
A	006	FI	4138	Malay apple, see Pomarac
A	013	VL	484	Mallow
A	001	FC	4011	Malta orange, see Blood Orange
B	031	MF	100	Mammalian fats (except milk fats)
A	006	FI	344	Mammey apple
A	001	FC	206	Mandarin, see also Subgroup 3 Mandarins
A	001	FC	3	Mandarins (including Mandarin-like hybrids)
C	052	AM	5255	Mangel or Mangold, see Fodder beet
A	014	VP	4435	Mangetout or Mangetout pea, see Podded pea

INDEX OF FOOD AND ANIMAL FEED COMMODITIES
(In Alphabetical order)

Class	Group			
A	006	FI	345	Mango
C	052	AM	5256	Mangoldwurzel, see Fodder beet
A	006	FI	346	Mangostan
A	006	FI	4137	Mangosteen, see Mangostan
A	014	VP	4435	Manila bean (immature pods), see Goa bean
A	016	VR	4555	Manioc, see Cassava, bitter
A	027	HH	737	Marigold flowers
B	049	IM	151	Marine bivalve molluscs
B	045	WC	145	Marine crustaceans
E	084	SC	145	Marine crustaceans, cooked
B	042	WS	125	Marine fish
E	080	MD	125	Marine fish, dried
B	044	WM	141	Marine mammals
A	027	HH	736	Marjoram
A	027	HH	4749	Marjoram, Sweet, see Marjoram
A	027	HH	4751	Marjoram, Wild, see Marjoram
D	057	DH	736	Marjoram, dry
A	006	FI	347	Marmaladedos
A	011	VC	4213	Marrow, see Squash, Summer
C	052	AV	1052	Marrow-stem cabbage or Marrow-stem kale
A	013	VL	471	Marsh marigold
A	015	VD	535	Mat bean (dry)
A	014	VP	535	Mat bean (green pods, mature, fresh seeds)
A	013	VL	4361	Matrimony vine, see Box thorn
D	066	DT	1113	Mat (dry leaves)
D	066	DT	5281	Mayweed, Scented, see Camomile, German
B	030	MM	95	Meat (from mammals other than marine mammals)
B	030	MM	96	Meat of cattle, goats, horses, pigs and sheep
B	030	MM	97	Meat of cattle, pigs and sheep
E	080	MD	95	Meat, dried (from mammals other than marine mammals)
A	001	FC	4014	Mediterranean mandarin, see Subgroup 3 Mandarins
B	041	WD	4915	Medium red salmon, see Subgroup Salmon, Pacific
A	002	FP	229	Medlar
C	050	AL	5223	Melilot, see Clovers
A	012	VO	4295	Melon pear, see Pepino
A	011	VC	4215	Melon, Crenshaw, see Subgroup Melons, except Watermelon
A	011	VC	4217	Melon, Honey Ball, see Subgroup Melons, except Watermelon
A	011	VC	4219	Melon, Honeydew, see Subgroup Melons, except Watermelon
A	011	VC	4221	Melon, Mango, see Subgroup Melons, except Watermelon
A	011	VC	4223	Melon, Netted, see Subgroup Melons, except Watermelon
A	011	VC	4225	Melon, Oriental Pickling, see Subgroup Melons, except Watermelon
A	011	VC	4227	Melon, Persian, see Subgroup Melons, except Watermelon
A	011	VC	4229	Melon, Pomegranate, see Subgroup Melons, except Watermelon
A	011	VC	4231	Melon, Serpent, see Subgroup Melons, except Watermelon
A	011	VC	4233	Melon, Snake, see Subgroup Melons, except Watermelon
A	011	VC	4235	Melon, White-skinned, see Subgroup Melons, except Watermelon
A	011	VC	4237	Melon, Winter, see Subgroup Melons, except Watermelon
A	011	VC	46	Melons, except Watermelon
B	042	WS	942	Menhaden
E	086	FM	183	Milk fats

INDEX OF FOOD AND ANIMAL FEED COMMODITIES
(In Alphabetical order)

Class	Group			
B	033	ML	107	Milk of cattle, goats and sheep
B	041	WD	891	Milkfish
B	033	ML	106	Milks
A	020	GC	646	Millet
C	051	AS	646	Millet fodder, dry
A	020	GC	4645	Millet, Barnyard, see Millet
A	020	GC	4647	Millet, Bulrush, see Millet
A	020	GC	4649	Millet, Common, see Millet
A	020	GC	4651	Millet, Finger, see Millet
A	020	GC	4653	Millet, Foxtail, see Millet
A	020	GC	4655	Millet, Little, see Millet
A	020	GC	4657	Milo, see Sorghum
C	052	AM	738	Mint hay
A	027	HH	738	Mints
D	057	DH	738	Mints, dry
A	003	FS	4057	Mirabelle, see Plum, Mirabelle
C	052	AM	165	Miscellaneous fodder and forage crops
B	049	IM	150	Molluscs, including Cephalopods
A	006	FI	348	Mombin, Yellow
B	030	MM	4815	Moose, European, meat, see Elk meat
A	003	FS	246	Morello
A	015	VD	4507	Moth bean (dry), see Mat bean (dry)
A	014	VP	4437	Moth bean, see Mat bean
B	030	MM	4813	Moufflon meat, see sheep meat
B	040	WF	4851	Mozambique tilapia, see Tilapia
A	027	HH	4753	Mugwort, see Wormwoods
A	004	FB	271	Mulberries
B	043	WR	943	Mullet roe (m)
B	042	WS	943	Mullets
A	009	VA	4167	Multiplying onion, see Onion, Welsh
A	015	VD	536	Mung bean (dry)
A	014	VP	536	Mung bean (green pods)
D	055	DF	5261	Muscatel, see Dried grapes
A	012	VO	450	Mushrooms
A	011	VC	4239	Muskmelon, see Subgroup Melons, except Watermelon
B	049	IM	1003	Mussels
A	013	VL	485	Mustard greens
A	023	SO	485	Mustard seed
A	023	SO	694	Mustard seed, Field
A	023	SO	478	Mustard seed, Indian
A	023	SO	90	Mustard seeds
A	013	VL	4364	Mustard spinach, see Komatsuma
A	013	VL	4363	Mustard, Indian, see Indian Mustard
A	003	FS	4059	Myrobolan plum, see Cherry plum
A	027	HH	4755	Myrrh, see Sweet Cicely
A	001	FC	4016	Myrtle-leaf orange, see Chinotto
A	013	VL	4365	Namenia, see Turnip greens
A	006	FI	349	Naranjilla
A	012	VO	4297	Naranjilla, see Group 6 Assorted tropical and sub-tropical fruits - inedible pee
A	002	FP	4047	Nashi pear, see Pear, oriental

INDEX OF FOOD AND ANIMAL FEED COMMODITIES
(In Alphabetical order)

Class	Group			
A	028	HS	739	Nasturtium pods
A	027	HH	739	Nasturtium, Garden, leaves
A	005	FT	304	Natal plum
A	001	FC	4018	Natsudaidai, see Subgroup 5 Shaddocks or Pomelos
A	015	VD	4509	Navy bean (dry), see Common bean (dry)
A	014	VP	4439	Navy bean (young pods and/or immature seeds),
A	003	FS	245	Nectarine
A	013	VL	486	New Zealand spinach
A	023	SO	695	Niger seed
A	013	VL	487	Nightshade, Black
B	041	WD	897	Nile perch
B	042	WS	4989	Northern bluefin tuna, see Subgroup Tuna and Bonito
B	040	WF	4849	Northern pike, see Pikes
A	028	HS	789	Nutmeg
C	051	AF	647	Oat forage (green)
C	051	AS	647	Oat straw and fodder, dry
A	020	GC	4659	Oat, Red, see Oats
A	020	GC	647	Oats
A	016	VR	586	Oca
B	042	WS	944	Ocean Perch
B	049	IM	5175	Octopus, Common, see Octopuses
B	049	IM	5177	Octopus, Curled, see Octopuses
B	049	IM	5179	Octopus, Musky, see Octopuses
B	049	IM	5173	Octopuses
B	042	WS	4991	Oil sardine, see Subgroup Sardines and Sardine type fishes
A	023	SO	88	Oilseed
A	023	SO	89	Oilseed except peanut
A	012	VO	442	Okra
A	004	FB	4087	Olallie berry, see Dewberries
D	068	OR	305	Olive oil, refined, as defined in Codex Stan. 33-1981
D	067	OC	305	Olive oil, virgin, see definition in Codex Stan. 33-1981
D	068	OR	5330	Olive, residue oil, as defined in Codex Stan. 33-1981, see Olive oil, refined
A	023	SO	4719	Olive, see Group 5: Assorted tropical and sub-tropical fruits - edible peel
A	005	FT	305	Olives
D	069	DM	305	Olives, processed
A	009	VA	385	Onion, Bulb
A	009	VA	386	Onion, Chinese
A	009	VA	4169	Onion, Egyptian, see Tree onion
A	009	VA	387	Onion, Welsh
A	013	VL	488	Orach
D	070	JF	4	Orange juice
A	001	FC	4019	Orange, Bitter, see Orange, Sour
A	001	FC	207	Orange, Sour, see also Subgroup 4 Oranges, Sweet, Sour
A	001	FC	208	Orange, Sweet, see also Subgroup 4 Oranges, Sweet, Sour
A	001	FC	4	Oranges, Sweet, Sour (including Orange-like hybrids): several cultivars
D	057	DH	5271	Oregano (= Wild Marjoram) dry, see Marjoram
A	027	HH	4757	Oregano, see Marjoram
A	005	FT	306	Otaheite gooseberry
A	016	VR	4557	Oyster plant, see Salsify
B	049	IM	5181	Oyster, American cupped, see Oysters

INDEX OF FOOD AND ANIMAL FEED COMMODITIES
(In Alphabetical order)

Class	Group			
B	049	IM	5183	Oyster, European, see Oysters
B	049	IM	5185	Oyster, Pacific cupped, see Oysters
B	049	IM	5187	Oyster, Portuguese cupped, see Oysters
B	049	IM	5189	Oyster, Sydney rock, see Oysters (including Cupped Oysters)
B	049	IM	1004	Oysters (including Cupped oysters)
A	022	TN	670	Pachira nut
B	041	WD	4917	Pacific salmon, see Subgroup Salmon, Pacific
B	041	WD	892	Paddle fish
A	013	VL	466	Pak-choi or Paksoi
A	013	VL	4367	Pak-tsai, see Chinese cabbage, (type Pe-tsai)
A	013	VL	4368	Pak-tsoi or Pak-soi, see Pak-choi or Paksoi
A	023	SO	696	Pala nut
A	017	VS	626	Palm hearts
D	067	OC	1240	Palm kernel oil, crude
D	068	OR	1240	Palm kernel oil, edible
D	067	OC	696	Palm oil, crude
D	068	OR	696	Palm oil, edible
A	006	FI	4139	Papaw or Pawpaw, see Papaya
A	006	FI	350	Papaya
A	013	VL	337	Papaya leaves
A	012	VO	4299	Paprika, see Peppers, Sweet
A	022	TN	671	Paradise nut, see Sapucaia nut
D	066	DT	5283	Paraguay tea, see Mat
A	027	HH	740	Parsley
A	016	VR	587	Parsley, Turnip-rooted
A	016	VR	588	Parsnip
B	036	PM	844	Partridge meat
A	006	FI	351	Passion fruit
A	011	VC	4241	Patisson, see Squash, White Bush
A	015	VD	4511	Pea (dry), see Field pea (dry)
C	050	AL	72	Pea hay or Pea fodder (dry)
C	050	AL	528	Pea vines (green)
A	014	VP	4441	Pea, see Garden pea
A	003	FS	247	Peach
A	023	SO	697	Peanut
C	050	AL	697	Peanut fodder
C	050	AL	1270	Peanut forage (green)
D	067	OC	697	Peanut oil, crude
D	068	OR	697	Peanut oil, edible
A	023	SO	703	Peanut, whole
A	002	FP	230	Pear
A	002	FP	4049	Pear, Oriental, see Pear
A	020	GC	4661	Pearl millet, see Millet, Bulrush
A	015	VD	72	Peas (dry)
A	014	VP	63	Peas (pods and succulent = immature seeds)
A	014	VP	64	Peas, shelled (succulent seeds)
A	022	TN	672	Pecan
A	027	HH	4759	Pennyroyal, see Mints
A	012	VO	443	Pepino
A	013	VL	489	Pepper leaves

INDEX OF FOOD AND ANIMAL FEED COMMODITIES
(In Alphabetical order)

Class	Group			
A	028	HS	790	Pepper, Black; White (see Note)
A	028	HS	791	Pepper, Long
D	066	DT	5285	Peppermint tea (succulent or dry leaves), see Peppermint, Group 027: Herbs
A	027	HH	4761	Peppermint, see Mints
A	012	VO	51	Peppers
A	012	VO	444	Peppers, Chili
A	012	VO	4301	Peppers, Long, see Peppers, Sweet
A	012	VO	445	Peppers, Sweet (including pimento or pimiento)
B	040	WF	864	Perch
B	040	WF	4853	Perch, American yellow, see Perch
B	040	WF	4855	Perch, European, see Perch
B	040	WF	870	Perch, Golden
B	040	WF	4857	Perch, White, see Perch
A	005	FT	4113	Persimmon Chinese, see Persimmon, Japanese
A	006	FI	352	Persimmon, American
A	005	FT	307	Persimmon, Japanese
A	006	FI	4141	Persimmon, Japanese, see Group 5
B	036	PM	845	Pheasant meat
B	031	MF	818	Pig fat
B	030	MM	818	Pig meat
E	080	MD	818	Pig meat, dried (including dried and smoked)
B	032	MO	818	Pig, Edible offal of
B	032	MO	1284	Pig, kidney
B	032	MO	1285	Pig, liver
A	014	VP	4443	Pigeon bean (green pods and immature seeds), see Broad bean
B	036	PM	846	Pigeon meat
A	015	VD	537	Pigeon pea (dry)
A	014	VP	537	Pigeon pea (green pods and/or young green seeds)
A	022	TN	4689	Pignolia or Pignoli, see Pine nuts
B	040	WF	865	Pike
B	040	WF	866	Pike-perch
A	022	TN	674	Pili nuts
A	012	VO	4303	Pimento or Pimiento, see Peppers, Sweet
A	028	HS	792	Pimento, fruit
A	022	TN	673	Pine nuts
A	006	FI	353	Pineapple
C	052	AM	353	Pineapple fodder
A	006	FI	4143	Pineapple guava, see Feijoa
D	070	JF	341	Pineapple juice
B	041	WD	4919	Pink salmon, see Subgroup Salmon, Pacific
A	022	TN	4691	Pinocchi, see Pine nuts
A	022	TN	675	Pistachio nut
A	005	FT	4115	Pitanga, see Surinam Cherry
A	022	TN	4693	PiÊon nut, see Pine nuts
B	042	WS	945	Plaice
B	042	WS	4993	Plaice, Alaska, see Plaice
B	042	WS	4995	Plaice, European, see Plaice
A	006	FI	354	Plantain
A	013	VL	490	Plantain leaves
A	003	FS	4061	Plum, American, see Sloe

INDEX OF FOOD AND ANIMAL FEED COMMODITIES
(In Alphabetical order)

Class	Group			
A	003	FS	248	Plum, Chickasaw
A	003	FS	4063	Plum, Damson, see Bullace
A	003	FS	4065	Plum, Greengage, see Plums
A	003	FS	4069	Plum, Japanese, see Plums
A	003	FS	4071	Plum, Mirabelle, see Bullace
A	003	FS	14	Plums (including Prunes)
A	014	VP	538	Podded pea (young pods)
A	013	VL	4369	Poke-berry leaves, see Pokeweed
A	013	VL	491	Pokeweed
B	042	WS	946	Pollack
A	005	FT	4119	Pomarrosa, Malay, see Pomerac
A	005	FT	4117	Pomarrosa, see Rose apples
A	002	FP	9	Pome fruits
A	006	FI	355	Pomegranate
A	001	FC	4020	Pomelo, see Shaddocks or Pomelos
A	005	FT	308	Pomerac
B	042	WS	947	Pomfret, Atlantic
A	020	GC	656	Popcorn
A	023	SO	698	Poppy seed
A	028	HS	4783	Poppy seed, see Group 023: Oilseed
B	042	WS	4997	Porbeagle, see Subgroup Sharks
B	044	WM	5051	Porpoise, see Whales
A	016	VR	589	Potato
A	016	VR	4559	Potato yam, see Yam bean
B	037	PF	111	Poultry fats
E	085	FA	111	Poultry fats, processed
B	036	PM	110	Poultry meat
B	038	PO	113	Poultry skin
B	038	PO	111	Poultry, Edible offal of
B	045	WC	5103	Prawn, Banana, see Shrimps or Prawns
B	045	WC	5105	Prawn, Brown tiger, see Shrimps or Prawns
B	045	WC	5107	Prawn, Caramote, see Shrimps or Prawns
B	045	WC	5109	Prawn, Common, see Shrimps or Prawns
B	045	WC	5111	Prawn, Eastern king, see Shrimps or Prawns
B	045	WC	5113	Prawn, Endeavour, see Shrimps or Prawns
B	045	WC	5115	Prawn, Giant tiger, see Shrimps or Prawns
B	045	WC	5117	Prawn, Green tiger, see Shrimps or Prawns
B	045	WC	5119	Prawn, Japanese king, see Shrimps or Prawns
B	045	WC	5121	Prawn, Kuruma, see Prawn, Japanese King
B	045	WC	5123	Prawn, Northern, see Shrimps or Prawns
B	045	WC	5125	Prawn, Western king, see Shrimps or Prawns
B	045	WC	5101	Prawns, see Shrimps or Prawns
A	006	FI	356	Prickly pear
E	085	FA	142	Processed Fat (Blubber), of Whales, Dolphins and Seals
A	020	GC	4665	Proso millet, see Millet, Common
D	055	DF	14	Prunes
A	003	FS	4072	Prunes, see Plums
C	050	AL	5227	Puero, see Kudzu, Tropical
A	006	FI	357	Pulasan
A	015	VD	70	Pulses

INDEX OF FOOD AND ANIMAL FEED COMMODITIES
(In Alphabetical order)

Class	Group			
A	011	VC	429	Pumpkins
A	013	VL	492	Purslane
A	013	VL	493	Purslane, Winter
B	039	PE	847	Quail eggs
B	036	PM	847	Quail meat
B	036	PM	4831	Quail, Bobwhite, see Quail
B	036	PM	4833	Quail, California, see Quail meat
A	022	TN	4695	Queensland Nut, see Macadamia nut
A	016	VR	4561	Queensland arrowroot, see Canna, edible
A	002	FP	231	Quince
A	020	GC	648	Quinoa
A	012	VO	4305	Quito Orange, see Naranjilla
A	006	FI	4145	Quito orange, see Naranjilla
B	031	MF	819	Rabbit fat
B	030	MM	819	Rabbit meat
A	016	VR	494	Radish
A	013	VL	494	Radish leaves (including Radish tops)
A	016	VR	590	Radish, Black
A	016	VR	591	Radish, Japanese
B	041	WD	4921	Rainbow trout, see Trout
D	055	DF	5263	Raisins (seedless white grape var., partially dried), see Dried grapes
A	009	VA	4171	Rakkyo, see Onion, Chinese
A	006	FI	358	Rambutan
A	016	VR	592	Rampion roots
A	013	VL	495	Rape greens
A	023	SO	495	Rape seed
D	067	OC	495	Rape seed oil, crude
D	068	OR	495	Rape seed oil, edible
A	023	SO	4721	Rape seed, Indian, see Mustard seed, Field
A	004	FB	272	Raspberries, Red, Black
B	042	WS	948	Rays
A	016	VR	4564	Red beet, see Beetroot
A	015	VD	4513	Red gram (dry), see Pigeon pea (dry)
A	014	VP	4447	Red gram (green pods and/or young green seeds), see Pigeon pea
A	013	VL	4371	Red leaved chicory, see Chicory leaves
B	030	MM	820	Reindeer meat
B	048	AR	149	Reptiles
B	042	WS	4999	Requiem shark, see Subgroup Sharks
B	040	WF	4859	Rhinofishes, see Carp, Indian
A	017	VS	627	Rhubarb
A	020	GC	649	Rice
A	015	VD	539	Rice bean (dry)
A	014	VP	539	Rice bean (young pods)
D	065	CF	649	Rice bran, processed
D	058	CM	1206	Rice bran, unprocessed
C	051	AS	649	Rice straw and fodder, dry
D	058	CM	649	Rice, husked
D	058	CM	1205	Rice, polished
B	040	WF	867	Roaches
B	045	WC	5127	Rock lobster, see Lobsters

INDEX OF FOOD AND ANIMAL FEED COMMODITIES
(In Alphabetical order)

Class	Group			
A	013	VL	4372	Rocket salad, see Rucola
B	030	MM	821	Roe meat
A	016	VR	75	Root and tuber vegetables
A	013	VL	4374	Roquette, see Rucola
A	005	FT	309	Rose apple
A	004	FB	273	Rose hips
A	012	VO	446	Roselle
D	066	DT	446	Roselle (calyx and flowers), dry
A	013	VL	446	Roselle leaves
A	027	HH	741	Rosemary
D	057	DH	741	Rosemary, dry
A	013	VL	496	Rucola
A	027	HH	742	Rue
D	057	DH	742	Rue, dry
A	014	VP	4449	Runner bean (green pods and seeds), see Common bean
A	015	VD	4515	Runner bean, see Common bean, Group 06: Legume vegetables
A	020	GC	4667	Russian millet, see Millet, Common
A	013	VL	497	Rutabaga greens
A	016	VR	4563	Rutabaga, see Swede
A	020	GC	650	Rye
D	065	CF	650	Rye bran, processed
D	058	CM	650	Rye bran, unprocessed
D	078	CP	1250	Rye bread
D	065	CF	1250	Rye flour
C	051	AF	650	Rye forage (green)
C	051	AS	650	Rye straw and fodder, dry
D	065	CF	1251	Rye wholemeal
A	023	SO	699	Safflower seed
D	067	OC	699	Safflower seed oil, crude
D	068	OR	699	Safflower seed oil, edible
A	027	HH	743	Sage and related Salvia species
D	057	DH	743	Sage, dry
C	050	AL	1027	Sainfoin
B	042	WS	5001	Salema, see Boque
B	043	WR	893	Salmon roe, Atlantic (d)
B	043	WR	121	Salmon roe, Pacific (d)
B	041	WD	893	Salmon, Atlantic, see Atlantic salmon
B	041	WD	121	Salmon, Pacific
B	041	WD	4923	Salmon, Pacific, see Subgroup Salmon, Pacific at the beginning of this group
B	042	WS	957	Salmon, Threadfin
A	016	VR	498	Salsify
A	013	VL	498	Salsify leaves
A	016	VR	4565	Salsify, Black, see Scorzonera
A	016	VR	593	Salsify, Spanish
A	002	FP	4051	Sand pear, see Pear, Oriental
A	006	FI	359	Sapodilla
A	006	FI	360	Sapote, Black
A	006	FI	361	Sapote, Green
A	006	FI	362	Sapote, Mammey
A	006	FI	363	Sapote, White

INDEX OF FOOD AND ANIMAL FEED COMMODITIES
(In Alphabetical order)

Class	Group			
A	022	TN	676	Sapucaia nut
B	042	WS	5005	Sardine, European
B	042	WS	5003	Sardinella or Oil sardine
B	042	WS	130	Sardines and Sardine-type fishes
A	027	HH	744	Sassafras leaves
A	001	FC	4022	Satsuma or Satsuma mandarin, see Subgroup 3 Mandarins
A	027	HH	745	Savory, Summer; Winter
D	057	DH	745	Savory, Summer; Winter, dry
B	042	WS	5007	Scad, see Jack Mackerel
B	049	IM	5191	Scallop, Australian, see Scallops
B	049	IM	5193	Scallop, Bay, see Scallops
B	049	IM	5195	Scallop, Giant Pacific, see Scallops
B	049	IM	5197	Scallop, Great, see Scallops
B	049	IM	5199	Scallop, New Zealand, see Scallops
B	049	IM	5201	Scallop, Queen, see Scallops
B	049	IM	5203	Scallop, Sea, see Scallops
B	049	IM	1005	Scallops
A	014	VP	540	Scarlet runner bean (pods and seeds)
A	015	VD	4517	Scarlet runner bean, see Group 06: Legume vegetables
B	042	WS	5009	Scorpion fishes, see Ocean Perch
A	016	VR	594	Scorzonera
B	042	WS	949	Sea bass
B	042	WS	950	Sea bream
B	041	WD	4925	Sea catfish, see Group 042: Marine fishes
B	042	WS	5011	Sea catfish, see Wolffish
A	005	FT	310	Sea grape
A	013	VL	499	Sea kale
B	049	IM	1006	Sea urchins
B	049	IM	1010	Sea-cucumbers
E	080	MD	1010	Sea-cucumbers, dried
B	044	WM	5053	Sea-lions, see Seals
B	044	WM	5055	Seal, Common, see Seals
B	044	WM	5063	Seal, Grey, see Seals
B	044	WM	5065	Seal, Harp, see Seals
B	044	WM	5067	Seal, Hooded, see Seals
B	044	WM	5069	Seal, Ringed, see Seals
B	044	WM	971	Seals
B	044	WM	5057	Seals, Eared, see Seals
B	044	WM	5059	Seals, Earless, see Seals
B	044	WM	5061	Seals, Fur, see Seals
A	024	SB	91	Seed for beverages
B	042	WS	5013	Seerfish, see Spanish Mackerel and King Mackerel
A	013	VL	500	Senna leaves
A	006	FI	364	Sentul
C	050	AL	5229	Sericea, see Lespedeza
A	004	FB	274	Service berries, see Juneberries
A	023	SO	700	Sesame seed
D	067	OC	700	Sesame seed oil, crude
D	068	OR	700	Sesame seed oil, edible
A	028	HS	4785	Sesame seed, see Group 023: Oilseed

INDEX OF FOOD AND ANIMAL FEED COMMODITIES
(In Alphabetical order)

Class	Group			
A	006	FI	4147	Sesso vegetal, see Akee apple
A	001	FC	4024	Seville Orange, see Orange, Sour
B	041	WD	894	Shad
B	043	WR	894	Shad roe (d)
A	001	FC	209	Shaddock, see also Subgroup 5 Shaddocks or Pomelos
A	001	FC	5	Shaddocks or Pomelos (including Shaddock-like hybrids, among others Grapefru
A	009	VA	388	Shallot
A	020	GC	4669	Shallu, see Sorghum
B	043	WL	131	Shark liver (m)
B	042	WS	5015	Shark, see Subgroup Sharks
B	042	WS	131	Sharks
A	023	SO	701	Shea nuts
B	031	MF	822	Sheep fat
B	030	MM	822	Sheep meat
B	033	ML	822	Sheep milk
E	086	FM	822	Sheep milk fat
E	085	FA	822	Sheep tallow
B	032	MO	822	Sheep, Edible offal of
B	032	MO	1288	Sheep, kidney
B	032	MO	1289	Sheep, liver
B	045	WC	5131	Shrimp, Deepwater rose, see Shrimps or Prawns
B	045	WC	5133	Shrimp, Northern brown, see Shrimps or Prawns
B	045	WC	5135	Shrimp, Northern pink, see Shrimps or Prawns
B	045	WC	5137	Shrimp, Northern white, see Shrimps or Prawns
B	045	WC	979	Shrimps or Prawns, (See Note 2)
E	084	SC	979	Shrimps or Prawns, cooked
E	084	SC	1220	Shrimps or Prawns, parboiled
B	045	WC	5129	Shrimps, Common, see Shrimps or Prawns
A	015	VD	4519	Sieva bean (dry), see Lima bean (dry)
A	014	VP	4451	Sieva bean (young pods and/or green fresh beans), see Lima bean
A	013	VL	4373	Silver beet, see Chard
B	041	WD	4927	Silver salmon, see Subgroup Salmon, Pacific
A	009	VA	390	Silverskin onion
A	011	VC	4243	Sinkwa or Sinkwa towel gourd, see Loofah, Angled
B	042	WS	5017	Skipjack tuna, see Subgroup Tuna and Bonito
A	016	VR	595	Skirrit or Skirret
B	045	WC	5139	Slipper lobster, see Lobsters
A	003	FS	249	Sloe
B	041	WD	895	Smelt
B	041	WD	4929	Smelt, European, see Smelt
B	041	WD	4931	Smelt, Rainbow, see Smelt
B	042	WS	5019	Smooth hounds, see Subgroup Sharks
B	049	IM	5205	Snail, Garden, see Snails, Edible
B	049	IM	5207	Snail, Giant, see Snails, Edible
B	049	IM	5209	Snail, Roman, see Snails, Edible
B	049	IM	1007	Snails, Edible
A	011	VC	430	Snake gourd
B	048	AR	992	Snakes
A	014	VP	4453	Snap bean (young pods), see Common bean
B	041	WD	4933	Sockeye salmon, see Subgroup Salmon, Pacific

INDEX OF FOOD AND ANIMAL FEED COMMODITIES
(In Alphabetical order)

Class	Group			
B	042	WS	951	Sole
A	020	GC	651	Sorghum
C	051	AF	651	Sorghum forage (green)
D	069	DM	658	Sorghum molasses
C	051	AS	651	Sorghum straw and fodder, dry
A	021	GS	658	Sorgo or Sorghum, Sweet
A	020	GC	4671	Sorgo, see Sorghum
A	027	HH	746	Sorrel, Common, and related Rumex species
A	013	VL	4388	Sorrel, Jamaican, see Roselle leaves
A	006	FI	365	Soursop
B	042	WS	5021	Southern bluefin tuna, see Tuna, Bluefin
A	027	HH	4763	Southernwood, see Wormwoods
A	013	VL	501	Sowthistle
A	015	VD	541	Soya bean (dry)
A	023	SO	4723	Soya bean (dry), see Group 07: Pulses
A	014	VP	541	Soya bean (immature seeds)
C	050	AL	541	Soya bean fodder
C	050	AL	1265	Soya bean forage (green)
D	067	OC	541	Soya bean oil, crude
D	068	OR	541	Soya bean oil, refined
A	015	VD	4521	Soybean (dry), see Soya bean (dry)
A	023	SO	4724	Soybean (dry), see Soya bean (dry)
A	014	VP	4455	Soybean, see Soya bean
A	006	FI	366	Spanish lime
B	042	WS	5023	Spanish mackerel, see King mackerel
A	027	HH	4765	Spearmint, see Mints
A	020	GC	4673	Spelt, see Wheat
A	028	HS	93	Spices
A	020	GC	4675	Spiked millet, see Millet, Bulrush
A	013	VL	502	Spinach
A	013	VL	4375	Spinach beet, see Chard
A	013	VL	503	Spinach, Indian
B	042	WS	5025	Spiny dogfish, see Subgroup Sharks
B	045	WC	5141	Spiny lobster, see Lobsters
A	011	VC	4245	Sponge gourd, see Loofah, Smooth
A	009	VA	389	Spring onion
B	041	WD	4935	Spring salmon, see Subgroup Salmon, Pacific
A	011	VC	431	Squash, Summer
A	011	VC	4249	Squash, White bush, see Squash, Summer
A	011	VC	4247	Squash, see Squash, Summer, and Winter Squash
B	049	IM	1009	Squid, Common, see Squids
B	049	IM	5211	Squid, European flying, see Squids
B	049	IM	5213	Squid, Japanese Flying, see Squids
B	049	IM	5215	Squid, Short-finned, see Squids
B	049	IM	1008	Squids
A	005	FT	4121	St. John's bread, see Carob
A	017	VS	78	Stalk and stem vegetables
A	006	FI	367	Star apple
E	080	MD	126	Stockfish (= dried Cod and Cod-like fishes)
A	003	FS	12	Stone fruits

INDEX OF FOOD AND ANIMAL FEED COMMODITIES
(In Alphabetical order)

Class	Group			
C	051	AS	81	Straw and fodder (dry) of cereal grains
C	051	AS	161	Straw, fodder (dry) and hay of cereal grains and other grass-like plants
A	004	FB	276	Strawberries, Wild
A	004	FB	275	Strawberry
A	006	FI	4149	Strawberry peach, see Kiwifruit
A	012	VO	4307	Strawberry tomato, see Ground cherries
A	004	FB	4091	Strawberry, Musky, see Strawberries, Wild
B	041	WD	896	Sturgeon
B	043	WR	896	Sturgeon roe (d)
A	006	FI	368	Sugar apple
A	016	VR	596	Sugar beet
D	069	DM	596	Sugar beet molasses
D	071	AB	596	Sugar beet pulp, dry
D	071	AB	1201	Sugar beet pulp, wet
A	021	GS	659	Sugar cane
C	052	AM	659	Sugar cane fodder
C	052	AV	659	Sugar cane forage
D	069	DM	659	Sugar cane molasses
A	013	VL	4377	Sugar loaf, see Chicory leaves
A	014	VP	4457	Sugar pea (young pods), see Podded pea
D	055	DF	5265	Sultanas, see Dried grapes
A	023	SO	702	Sunflower seed
D	067	OC	702	Sunflower seed oil, crude
D	068	OR	702	Sunflower seed oil, edible
A	005	FT	311	Surinam cherry
A	016	VR	497	Swede
C	052	AM	497	Swedish turnip or Swede fodder
A	027	HH	747	Sweet Cicely
D	057	DH	747	Sweet Cicely, dry
A	012	VO	447	Sweet corn (corn-on-the-cob), see definition in Codex Stan. 133-1981
A	012	VO	1275	Sweet corn (kernels), see definition in Codex Stan. 132-1981
A	016	VR	508	Sweet potato
A	013	VL	508	Sweet potato, leaves
A	006	FI	4151	Sweetsop, see Sugar apple
A	013	VL	4379	Swiss chard, see Chard
A	014	VP	542	Sword bean (young pods and beans)
A	004	FB	1235	Table-grapes
B	042	WS	5027	Tailor (Australia), see Bluefish
E	085	FA	96	Tallow and lard from cattle, goats, horses, pigs and sheep
A	005	FT	4123	Tamarillo, see Tree tomato
A	006	FI	369	Tamarind
A	028	HS	4787	Tamarind, see Group 6: Assorted tropical and sub-tropical fruits - inedible pe...
A	001	FC	4029	Tangelo, large-sized cultivars, see Subgroup 5 Shaddocks or Pomelos
A	001	FC	4031	Tangelo, small and medium sized cultivars, see Subgroup 3 Mandarins
A	001	FC	4033	Tangelolo, see Subgroup 5 Shaddocks or Pomelos
A	001	FC	4027	Tangerine, see Subgroup 3 Mandarins
A	001	FC	4035	Tangors, see Subgroup 3 Mandarins
A	016	VR	4567	Tanier, see Tannia
A	001	FC	4037	Tankan mandarin, see Subgroup 3 Mandarins
A	016	VR	504	Tannia

INDEX OF FOOD AND ANIMAL FEED COMMODITIES
(In Alphabetical order)

Class	Group			
A	013	VL	504	Tannia leaves
A	027	HH	748	Tansy and related species
D	057	DH	748	Tansy and related species, dry
A	016	VR	4569	Tapioca, see Cassava
A	016	VR	505	Taro
A	013	VL	505	Taro leaves
A	027	HH	749	Tarragon
D	066	DT	1114	Tea, Green, Black (black, fermented and dried)
D	066	DT	171	Teas (Tea and Herb teas)
A	020	GC	652	Teff or Tef
A	013	VL	4381	Tendergreen, see Turnip greens
A	020	GC	657	Teosinte
C	051	AS	657	Teosinte fodder
A	015	VD	564	Tepary bean (dry)
D	057	DH	750	Thyme, dry
A	016	VR	580	Tiger nut
B	040	WF	868	Tilapia
A	012	VO	4309	Tomatillo, see Ground cherries
A	012	VO	448	Tomato
D	070	JF	448	Tomato juice
A	006	FI	370	Tonka bean
A	028	HS	370	Tonka bean, see also Group 6: Assorted tropical and sub-tropical fruits - inedibl
A	016	VR	589	Topee tambu
A	012	VO	4311	Tree melon, see Pepino
A	022	TN	85	Tree nuts
A	009	VA	391	Tree onion
A	005	FT	4125	Tree strawberry, see Arbutus berry
A	005	FT	312	Tree tomato
C	050	AL	1028	Trefoil
A	020	GC	653	Triticale
A	022	TN	677	Tropical almond
C	050	AL	5231	Tropical kudzu, see Kudzu, Tropical
B	041	WD	123	Trout
A	013	VL	4383	Tsai shim, see Choisum
A	013	VL	4385	Tsoi sum, see Choisum
B	042	WS	132	Tuna and Bonito
B	042	WS	5029	Tuna, Bigeye, see Subgroup Tuna and Bonito
B	042	WS	5031	Tuna, Blackfin, see Subgroup Tuna and Bonito
B	042	WS	5033	Tuna, Bluefin, see Subgroup Tuna and Bonito
B	042	WS	5035	Tuna, Longtail, see Subgroup Tuna and Bonito
B	042	WS	5037	Tuna, Skipjack, see Subgroup Tuna and Bonito
B	042	WS	5039	Tuna, Yellowfin, see Subgroup Tuna and Bonito
B	042	WS	952	Tuna, see also Subgroup Tuna and Bonito
B	042	WS	953	Turbot
E	085	FA	848	Turkey Fat, processed
B	037	PF	848	Turkey fat
B	036	PM	848	Turkey meat
B	038	PO	848	Turkey, Edible offal of
A	028	HS	794	Turmeric, root
C	052	AM	506	Turnip fodder

INDEX OF FOOD AND ANIMAL FEED COMMODITIES
(In Alphabetical order)

Class	Group			
A	013	VL	506	Turnip greens
C	052	AV	506	Turnip leaves or tops
A	016	VR	506	Turnip, Garden
A	016	VR	4573	Turnip, Swedish, see Swede
A	016	VR	4571	Turnip, see Swede
B	048	AR	5157	Turtle, Green, see Turtles
B	048	AR	5159	Turtle, Hawksbill, see Turtles
B	048	AR	5161	Turtle, Loggerhead, see Turtles
B	048	AR	993	Turtles
A	001	FC	4039	Ugli, see Subgroup 5 Shaddocks or Pomelos, Tangelo
A	016	VR	599	Ullucu
A	015	VD	4523	Urd bean (dry), see Black gram (dry)
A	014	VP	4459	Urd bean (green pods), see Black gram
A	004	FB	19	Vaccinium berries, including Bearberry
A	028	HS	795	Vanilla, beans
B	030	MM	4817	Veal (= calf meat), see Cattle meat
D	067	OC	172	Vegetable oils, crude
D	068	OR	172	Vegetable oils, edible
A	011	VC	4251	Vegetable spaghetti, see Pumpkins
A	011	VC	4253	Vegetable sponge, see Loofah, Smooth
C	050	AL	5233	Velvet bean, see Bean, Velvet
C	050	AL	1029	Vetch
C	050	AL	5235	Vetch, Chickling, see Vetch
C	050	AL	5237	Vetch, Crown, see Vetch
C	050	AL	5239	Vetch, Milk, see Vetch
A	013	VL	4387	Vine spinach, see Spinach, Indian
A	022	TN	4697	Walnut, Black, see Walnuts
A	022	TN	4699	Walnut, English; Walnut, Persian, see Walnuts
A	022	TN	678	Walnuts
B	030	MM	4819	Water Buffalo, meat, see Buffalo meat
A	013	VL	4389	Water Spinach, see Kangkung
A	013	VL	473	Watercress
A	027	HH	4767	Watercress, see Group 05: Leafy vegetables
A	011	VC	432	Watermelon
A	011	VC	4255	Wax gourd
A	011	VC	4257	West Indian gherkin, see Gherkin, West Indian
B	044	WM	5073	Whale, Blue, see Whales
B	044	WM	5075	Whale, False killer, see Whales
B	044	WM	5077	Whale, Fin, see Whales
B	044	WM	5079	Whale, Humpback, see Whales
B	044	WM	5081	Whale, Killer, see Whales
B	044	WM	5083	Whale, Minke, see Whales
B	044	WM	5085	Whale, Sei, see Whales
B	044	WM	5087	Whale, Short-finned pilot, see Whales
B	044	WM	5089	Whale, Sperm, see Whales
B	044	WM	972	Whales
B	044	WM	5071	Whales, Baleen, see Whales
E	085	FA	972	Whales, Blubber of, processed
B	044	WM	5091	Whales, Toothed, see Whales
A	020	GC	654	Wheat

INDEX OF FOOD AND ANIMAL FEED COMMODITIES
(In Alphabetical order)

Class	Group			
D	065	CF	654	Wheat bran, processed
D	058	CM	654	Wheat bran, unprocessed
D	065	CF	1211	Wheat flour
D	065	CF	1210	Wheat germ
C	051	AS	654	Wheat straw and fodder, dry
D	065	CF	1212	Wheat wholemeal
D	078	CP	1211	White bread
B	040	WF	4863	White crappie, see Perch, white
B	040	WF	4861	White perch, see Perch, white
B	042	WS	953	Whiting
D	078	CP	1212	Wholemeal bread
A	004	FB	4093	Whortleberry, Red, see Bilberry, Red
B	030	MM	823	Wild boar meat
A	020	GC	655	Wild rice
A	001	FC	4041	Willowleaf mandarin, see Mediterranean Mandarin and Subgroup 3 Mandarins
A	004	FB	1236	Wine-grapes
A	014	VP	4461	Winged bean (immature pods), see Goa bean
A	014	VP	543	Winged pea (young pods)
A	027	HH	751	Winter cress, Common; American
A	011	VC	4259	Winter melon, see Melon, Winter
A	011	VC	433	Winter squash, see also Pumpkins
A	027	HH	752	Wintergreen leaves
D	057	DH	752	Wintergreen leaves, dry
B	042	WS	5041	Witch flounder, see Flounders
A	017	VS	469	Witloof chicory (sprouts)
B	042	WS	955	Wolffish
A	027	HH	753	Woodruff
D	057	DH	753	Woodruff, dry
A	027	HH	754	Wormwoods
D	057	DH	754	Wormwoods, dry
A	015	VD	4525	Wrinkled pea (dry), see Field pea (dry)
A	014	VP	4463	Wrinkled pea, see Garden pea
B	030	MM	4821	Yak meat, see Cattle meat
A	016	VR	601	Yam bean
A	016	VR	4575	Yam, Cush-cush, see Yams
A	016	VR	4577	Yam, Eight-months, see Yam, White Guinea
A	016	VR	4579	Yam, Greater, see Yams
A	016	VR	4583	Yam, Twelve-months, see Yam, Yellow Guinea
A	016	VR	4587	Yam, White Guinea, see Yams
A	016	VR	4585	Yam, White, see Yam, White Guinea
A	016	VR	4591	Yam, Yellow Guinea, see Yams
A	016	VR	4589	Yam, Yellow, see Yam, Yellow Guinea
A	016	VR	600	Yams
A	014	VP	544	Yard-long bean (pods)
A	013	VL	4391	Yautia leaves, see Tannia leaves
A	016	VR	4593	Yautia, see Tannia
B	042	WS	5043	Yellowfin tuna, see Tuna, Yellowfin
B	042	WS	5044	Yellowtail flounder, see Flounders
A	004	FB	4094	Youngberry, see Dewberries
B	030	MM	4823	Zebu meat, see Cattle meat
A	011	VC	4261	Zucchetti, see Squash, Summer
A	011	VC	4263	Zucchini, see Squash, Summer

CLASS A PRIMARY FOOD COMMODITIES OF PLANT ORIGIN

TYPE 1   FRUITS

Fruits are derived from many different kinds of perennial plants, trees and shrubs, usually cultivated. They consist mostly of the ripe, often sweet, succulent or pulpy developed plant ovary and its accessory parts, commonly and traditionally known as fruit.

Exposure to pesticides is dependent on the particular part of the fruit used for food. Fruits may be consumed whole, after removal of the peel, or in part, and in the form of fresh, dried or processed products.

Citrus fruits (except kumquats)

Class A

Type 1   Fruits Group 1 Group Letter Code FC

Kumquats: See Group 5 Assorted tropical and sub-tropical fruits - edible peel

Citrus fruits are produced on trees or shrubs of the family *Rutaceae*. These fruits are characterized by aromatic oily peel, globular form and interior segments of juice-filled vesicles. The fruit is fully exposed to pesticides during the growing season. Post-harvest treatments with pesticides and liquid waxes are often carried out to avoid deterioration during transport and distribution due to fungal diseases, insect pests or loss of moisture.

The fruit pulp may be consumed in succulent form and as a juice. The entire fruit may be used for preserves.

Portion of the commodity to which the MRL applies (and which is analyzed): *Whole commodity.*

Group 1   **Citrus fruits**

<u>**Code No**</u>:   <u>**Commodity**</u>

FC   1   **Citrus fruits**

FC   2   **Lemons and Limes** (including Citron)
            - *Citrus limon* Burm.f.:
            - C. *aurantifolia* Swingle:
            - C. *medica* L.;
            Hybrids and related species similar to lemons and limes including Citrus
            jambhiri Lush.: C. *limetta Rosso*; C. *limettoides* Tan.: C. *limonia* Osbeck.
            synonyms: see specific fruit species

FC   3   **Mandarins** (including Mandarin-like hybrids)
            - *Citrus reticulata blanco*:
            Hybrids and related species including *C.nobilis* Lour.: *C. deliciosa* Ten.; C.
            tangarina Hort.: C. *mitis blanco*
            syn: C. *madurensis* Lour.:
            C. unshiu Markovitch
            synonyms: see specific fruit species of Mandarin

FC   4   **Oranges, Sweet, Sour** (including Orange-like hybrids): several cultivars
            - C. *sinensis* Osbeck:
            - C. *aurantium* L.:
            Hybrids and related species: *Citrus myrtifolia* Raf.: C. *salicifolia* Raf.:
            synonyms: see specific fruit species

FC   5   **Shaddocks or Pomelos** (including Shaddock-like hybrids, among others Grapefruit)
            - C. *Grandis* (L.) Osbeck:
            - C. *paradisi* Macf.:
            Hybrids and related species, similar to Shaddocks, including C. natsudaidai
            Hayata: Tangelos large sized (= hybrid of Grapefruit X Mandarin): Tangelolos:
            (hybrid of Grapefruit X Tangelo);
            synonyms: see specific fruit species

FC 4000   **Bigarade**, see Orange, Sour

FC 4001   **Blood orange**, see Orange, Sweet
            Cultivar of *Citrus sinensis* Osbeck

FC 201   **Calamondin**, see also Subgroup 3 Mandarins
              Citrus mitis blanco;
              syn: C. *madurensis* Lour. (hybrid of C. *reticulata blanco*, var. *austera* Swing
              X *Fortunella* sp.

FC 4002   **Chinotto**, see Orange, Sour
              Citrus aurantium L., var. *myrtifolia* Ker-Gawler:
              syn: C. *myrtifolia* Raf.

FC 4003   **Chironja**, see Subgroup Oranges, Sweet, Sour (including Orange-like hybrids)
                  = hybrid of Orange, Sweet X Mandarin

FC 202   **Citron**, see also Subgroup 2 Lemons and Limes
              Citrus medica L.:
              syn: C. *cedra* Link: C. *cedratus* Raf.: C. *medica genuina* Engl.: C. *medica*
              proper Bonavia

FC 4005   **Clementine**, see Mandarins
              Citrus clementina Hort. ex Tanaka cultivar of C. *reticulata blanco* (possibly
              natural hybrid of Mandarin X Orange, Sweet)

FC 4006   **Cleopatra mandarin**, see Subgroup 3 Mandarins
              C. *reshni* Hort. ex Tan.

FC 4007   **Dancy or Dancy mandarin**, see Subgroup 3 Mandarins
              C. *tangerina* Hort.

FC 203   **Grapefruit**, see also Subgroup 5 Shaddocks or Pomelos
              Hybrid of Shaddock X Orange, Sweet
              Citrus paradisi Macf.:
              syn: C. *maxima uvacarpa* Merr. & Lee.

FC 4008   **King mandarin**, see Subgroup 3 Mandarin
              C. *nobilis* Lour. (= hybrid of Mandarin X Orange, Sweet)

FC 204   **Lemon**, see also Subgroup 2 Lemons and Limes
              Citrus limon Burm.f.:
              syn: C. *medica limon* L.: C. *limonum Risso*: C. *medica limonum* Hook. f.

FC 205   **Lime**, see also Subgroup 2 Lemons and Limes
              Citrus aurantifolia Swingle;
              syn: *Limonia aurantifolia* Christm.: L. *acidissima* Houtt. *Citrus lima* Lunan.
              C. *acida* Roxb.: C. *limonellus* Hassk.

FC 4011   **Malta orange**, see Blood Orange

FC 206   **Mandarin**, see also Subgroup 3 Mandarins
            Citrus reticulata Blanco:
            syn: C. *nobilis* Andrews (non Lour.): C. *poonensis* Hort. ex Tanaka C. *chrysocarpa* Lush.

FC 4014   **Mediterranean** mandarin, see Subgroup 3 Mandarins
            Citrus deliciosa Ten (= hybrid of Mandarin X Orange, Sweet)

FC 4016   **Myrtle-leaf orange**, see Chinotto

FC 4018   **Natsudaidai**, see Subgroup 5 Shaddocks or Pomelos
            C. natsudaidai Hayata (possibly natural hybrid of Mandarin X Shaddock)

FC 4019   **Orange, Bitter**, see Orange, Sour

FC 207   **Orange, Sour**, see also Subgroup 4 Oranges, Sweet, Sour
            Citrus aurantium L.:
            syn: C. *vulgaris Risso*: C. *bigarradia* Loisel. C. *communis* Le Maout & Dec.

FC 208   **Orange, Sweet**, see also Subgroup 4 Oranges, Sweet, Sour
            Citrus sinensis Osbeck:
            syn: C. *aurantium sinensis* L.: C. *dulcis* Pers.: C. *aurantium vulgare Risso* & Poit.: C. *aurantium dulce* Hayne

FC 4020   **Pomelo**, see Shaddocks or Pomelos

FC 4022   **Satsuma or Satsuma mandarin**, see Subgroup 3 Mandarins
            Citrus unshiu Markovitch.

FC 4024   **Seville Orange**, see Orange, Sour

FC 209   **Shaddock**, see also Subgroup 5 Shaddocks or Pomelos
            Citrus grandis osbeck;
            syn: C. *aurantium decumana* L.: C. *maxima* (Burm.) Merr.: C. *decumana* Murr.

FC 4029   **Tangelo, large-sized cultivars**, see Subgroup 5 Shaddocks or Pomelos

FC 4031   **Tangelo, small and medium sized cultivars**, see Subgroup 3 Mandarins
            Hybrids of Mandarin X Grapefruit or Mandarin X Shaddock

FC 4033   **Tangelolo**, see Subgroup 5 Shaddocks or Pomelos
              Hybrids of Grapefruit X Tangelo

FC 4027   **Tangerine**, see Subgroup 3 Mandarins
              Citrus tangarina Hort. ex Tan.

FC 4035   **Tangors**, see Subgroup 3 Mandarins
              Citrus nobilis Lour. (= Hybrid of Mandarin X Orange, Sweet)

FC 4037   **Tankan mandarin**, see Subgroup 3 Mandarins
              Citrus tankan Hyata (= probably hybrid of Mandarin X Orange, Sweet)

FC 4039   **Ugli**, see Subgroup 5 Shaddocks or Pomelos, Tangelo,
              see large sized fruit cultivar

FC 4041   **Willowleaf mandarin**, see Mediterranean Mandarin and Subgroup 3 Mandarins
              Citrus deliciosa Ten. (= hybrid of Mandarin and Orange, Sweet)

Pome fruits

<u>Class A</u>
Type 1   **Fruits   Group 2   Group Letter Code FP**

Pome fruits are produced on trees and shrubs belonging to certain genera of the rose family (*Rosaceae*), especially the genera *Malus* and *Pyrus*. They are characterized by fleshy tissue surrounding a core consisting of parchment-like carpels enclosing the seeds.

Pome fruits are fully exposed to pesticides applied during the growing season. Post-harvest treatments directly after harvest may also occur. The entire fruit, except the core, may be consumed in the succulent form or after processing.

<u>Portion of the commodity to which the MRL applies (and which is analyzed)</u>: *Whole commodity after removal of stems.*

Group 2   **Pome fruits**

<u>**Code No.**</u>   <u>**Commodity**</u>

FP   9   **Pome fruits**

FP 226   **Apple**
         Malus domesticus Borkhausen

FP 227   **Crab-apple**
         Malus spp.; among others *Malus baccata* (L.) Rorkh.; M. *prunifolia* (Willd.) Borkh.

FP 4044   **Japanese medlar**, see Loquat

FP 228   **Loquat**
         Eriobotrya japonica (Thunberg ex J.A. Murray) Lindley

FP 229   **Medlar**
         Mespilus germanica L.

FP 4047   **Nashi pear**, see Pear, oriental

FP 230   **Pear**
         Pyrus communis L.; P. *pyrifolia* (Burm.) Nakai:
         P. bretschneideri Rhd.: P. *sinensis* L.

FP 4049   **Pear, Oriental**, see Pear
                   Pyrus pyritolia (Burm.) Nakai

FP 231   **Quince**
                   Cydonia oblonga P. Miller:
                   syn: *Cydonia vulgaris* Persoon

FP 4051   **Sand pear**, see Pear, Oriental

Stone fruits

<u>Class A</u>
Type 1   Fruits Group 3 Group Letter Code FS

Stone fruits are produced on trees belonging to the *genus Prunus* of the rose family (*Rosaceae*). They are characterized by fleshy tissue surrounding a single hard shelled seed. The fruit is fully exposed to pesticides applied during the growing season (from fruit setting until harvest). Dipping of fruit immediately after harvest, especially with fungicides, may also occur.

The entire fruit, except the seed, may be consumed in a succulent or processed form.

<u>Portion of the commodity to which the MRL applies (and which is analyzed)</u>: *Whole commodity after removal of stems and stones, but the residue calculated and expressed on the whole commodity without stem.*

Group 3   Stone fruits

<u>Code No.</u>   <u>Commodity</u>

FS   12   **Stone fruits**
             Prunus spp.

FS   13   **Cherries**
             Prunus cerasus L.: *P. avium* L.

FS   14   **Plums (including Prunes)**
             Prunus domestica L.: other *Prunus* spp and ssp.

FS 240   **Apricot**
             Prunus armeniaca L.:
             syn: *Armeniaca vulgaris* Lamarck

FS 241   **Bullace**
             Prunus insititia L.;
             syn: *Prunus domestica* L., ssp. *insititia* (L.) Schneider

FS 242   **Cherry plum**
             Prunus cerasifera Ehrhart, syn: *P. divaricata* Ledeboer *P. salicina* Lindl., var.
             Burbank

FS 4053   **Chickasaw plum**, see Plum, Chickasaw

FS 243   **Cherry, Sour**
             Prunus cerasus L.

FS 244   **Cherry, Sweet**
             Prunus avium L.

FS 4055   **Damsons (Damson plums)** see Plum, Damson

FS 4056   **Greengages (Greengage plums)**, see Plum, Greengage

FS 245   **Nectarine**
             Prunus persica (L.) Batch, var. *nectarina*

FS 4057   **Mirabelle**, see Plum, Mirabelle

FS 246   **Morello**
             Prunus cerasus L., var. *austera* L.

FS 4059   **Myrobolan plum**, see Cherry plum

FS 247   **Peach**
             Prunus persica (L.) Batsch:
             syn: P. *vulgaris* Mill.

FS 4061   **Plum, American**, see Sloe

FS 248   **Plum, Chickasaw**
             Prunus angustifolia Marsh.:
             syn: P. Chicasaw Mich.

FS 4063   **Plum, Damson**, see Bullace

FS 4065   **Plum, Greengage**, see Plums
             Prunus insititia L., var. *italica* (Borkh.) L.M. Neum.

FS 4069   **Plum, Japanese**, see Plums
             Prunus salicina Lindley:
             syn: P. *triflora* Roxb.

FS 4071   **Plum, Mirabelle**, see Bullace
             Prunus insititia L., var. *syriaca*:
             syn: P. *domestica* L., ssp *insititia* (L.) Schneider

FS 4072   **Prunes**, see Plums

FS 249   **Sloe**

Prunus spinosa L.; several wild *Prunus* spp.

Berries and other small fruits

Class A
Type 1   Fruits Group 4 Group Letter Code FB

Berries and other small fruits are derived from a variety of perennial plants and shrubs having fruit characterized by a high surface: weight ratio. The fruits are fully exposed to pesticides applied during the growing season (blossoming until harvest).

The entire fruit, often including seed, may be consumed in a succulent or processed form.

Portion of commodity to which the MRL applies (and which is analyzed): *Whole commodity after removal of caps and stems.Currants, Black, Red, White: fruit with stem.*

Group 4   Berries and other small fruits

Code No.	Commodity
FB 18	**Berries and other small fruits**
FB 19	**Vaccinium berries, including Bearberry** *Vaccinium* spp.: *Arctostaphylos uva-ursi* (L.) Spreng.
FB 20	**Blueberries** *Vaccinium corymbosum* L. *Vaccinium angustifolium* Ait.: *Vaccinium* ashei Reade; *Gaylussacia* spp.
FB 260	**Bearberry** *Arctostaphylos uva-ursi* (L.) Spreng.
FB 261	**Bilberry** *Vaccinium myrtillus* L.
FB 262	**Bilberry, Bog** *Vaccinium uliginosum* L.
FB 263	**Bilberry, Red** *Vaccinium vitis-idaea* L.
FB 264	**Blackberries** *Rubus fruticosus* L., several ssp.
FB 4073	**Blueberry, Highbush**, see Blueberries *Vaccinium corymbosum* L.

FB 4075   **Blueberry, Lowbush**, see Blueberries
                    Vaccinium angustifolium Ait.

FB 4077   **Blueberry, Rabbiteye**, see Blueberries
                    Vaccinium ashei Reade

FB 4079   **Boysenberry**, see Dewberries
                    Hybrid of *Rubus* spp.

FB 277   **Cloudberry**
                    Rubus chamaemorus L.

FB 4081   **Cowberry**, see Bilberry, Red

FB 265   **Cranberry**
                    Vaccinium macrocarpon Ait.
                    syn: *Oxycoccus macrocarpus* (Aiton) Pursh

FB 21   **Currants, Black, Red, White**
                    Ribes nigrum L.: R *rubrum* L.

FB 278   **Currant, Black**, see also Currants, Black, Red, White
                    Ribes nigrum L.

FB 279   **Currant, Red, White**, see also Currants, Black, Red, White
                    Ribes rubrum L.

FB 266   **Dewberries** (including Boysenberry and Loganberry)
                    Rubus ceasius L.: several *Rubus* ssp. and hybrids

FB 267   **Elderberries**
                    Sambucus spp.

FB 268   **Gooseberry**
                    Ribes uva-crispa L.
                    syn: *Ribes grossularia* L.

FB 269   **Grapes**
                    Vitis vinifera L., several cultivars

FB 4083   **Huckleberries**
                  1. Blueberries, see above
                  2. *Gaylussacia* spp., see Blueberries

FB 270   **Juneberries**
            Amelanchier *ovalis* Med.; A. Canadense Med.

FB 4085   **Loganberry**, see Dewberries
            Rubus loganobaccus, hybrid of *Rubus* spp.

FB 271   **Mulberries**
            Morus alba L.; *Morus nigra* L.: *Morus rubra* L.

FB 4087   **Olallie berry**, see Dewberries

FB 272   **Raspberries, Red, Black**
            Rubus idaeus L.; *Rubus occidentalis* L.

FB 273   **Rose hips**
            Rosa L., several spp.

FB 274   **Service berries**, see Juneberries
            Sorbus torminalis (L.) Crantz: *Sorbus domestica* L.

FB 275   **Strawberry**
            Fragaria x *ananassa* Duchene;
            syn: F. *grandiflora* Ehrh.

FB 276   **Strawberries, Wild**
            Fragaria vesca L., *Fragaria moschata* Duchene

FB 4091   **Strawberry, Musky**, see Strawberries, Wild
            Fragaria moschata Duchene

FB 1235   **Table-grapes**
            Special cultivars of *Vitis vinifera* L., suitable for direct human consumption

FB 4093   **Whortleberry, Red**, see Bilberry, Red

FB 4094   **Youngberry**, see Dewberries

FB 1236   **Wine-grapes**
            Special cultivars of *Vitis vinifera* L. suitable for preparing juice and fermenting
            into wine

Assorted tropical and sub-tropical fruits - edible peel

<u>Class A</u>
Type 1   Fruits Group 5 Group Letter Code FT

The assorted tropical and sub-tropical fruits - edible peel are derived from the immature or mature fruits of a large variety of perennial plants, usually shrubs or trees. The fruits are fully exposed to pesticides during the growing season (period of fruit development).

The whole fruit may be consumed in a succulent or processed form.

<u>Portion of the commodity to which the MRL applies (and which is analyzed)</u>: *Whole commodity. Dates and Olives: Whole commodity after removal of stems and stones but residue calculated and expressed on the whole fruit.*

Group 5   Assorted tropical and sub-tropical fruits - edible peel

<u>Code No.</u>   <u>Commodity</u>

FT   26   **Assorted tropical and sub-tropical fruits - edible peel**

FT 4095   **Acerola**, see Barbados cherry

FT 285   **Ambarella**
                Spondias cytherea Sonn.;
                syn: S. *dulcis* Forst

FT 4097   **Aonla**, see Otaheite gooseberry

FT 286   **Arbutus berry**
                Arbutus unedo L.

FT 287   **Barbados cherry**
                Malpighia glabra L.

FT 288   **Bilimbi**
                Averrhoa Bilimbi L.

FT 4099   **Brazilian cherry**, see Grumichana

FT 289   **Carambola**
                Averrhoa carambola L.

FT 290   **Caranda**
           Carissa carandas L.

FT 291   **Carob**
           Ceratonia siliqua L.

FT 292   **Cashew apple**
           Anacardium occidentale L.

FT 293   **Chinese olive, Black, White**
           Canarium pimela Koenig; syn: *C. nigrum* Engl.:
           Canarium album (Lour.) Taeusch.

FT 294   **Coco plum**
           Chrysobalanus icaco L.

FT 295   **Date**
           Phoenix dactylifera L.

FT 296   **Desert date**
           Balanites aegiptica Del.

FT 297   **Fig**
           Ficus carica L.

FT 298   **Grumichama**
           Eugenia dombeyana DC.

FT 299   **Hog plum**
           Spondias mombin L.:
           syn: S. *lutea* L.

FT 4101   **Icaco plum,** see Coco plum

FT 300   **Jaboticaba**
           Myrciaria cauliflora Berg.:
           syn: *Eugenia cauliflora* (Berg.) DC.

FT 4103   **Java almond,** see Group 024: Tree nuts

FT 301   **Jujube, Indian**
           Zizyphus mauritania Lam.:
           syn: Z. *jujuba* (L.) Lam. non Mill.

FT 302   **Jujube, Chinese**
         Zizyphus jujuba Mill.

FT 4105   **Kaki or Kaki fruit**, see Persimmon, Japanese

FT 303   **Kumquats**
         Fortunella japonica (Thunberg) Swingle:
         F. *margarita* (Loureiro) Swingle

FT 4107   **Kumquat, Marumi**, see Kumquats
         Fortunella japonica (Thunberg) Swingle

FT 4109   **Kumquat, Nagami**, see Kumquats
         Fortunella margarita (Loureiro) Swingle

FT 4111   **Locust tree**, see Carob

FT 304   **Natal plum**
         Carissa grandiflora A.DC.

FT 305   **Olives**
         Olea europaea L., *var. europaea*

FT 306   **Otaheite gooseberry**
         Phyllantus distichus (L.) Muell.- Arg.
         syn: Ph. *acidus* (L.) Skeels

FT 4113   **Persimmon Chinese**, see Persimmon, Japanese

FT 307   **Persimmon, Japanese**
         Diospyros Kaki L.f.:
         syn: D. *chinensis* Blume
         some cultivars have an inedible peel

FT 4115   **Pitanga**, see Surinam Cherry

FT 4117   **Pomarrosa**, see Rose apples

FT 4119   **Pomarrosa, Malay**, see Pomerac

FT 308   **Pomerac**
         Syzygium Malaccensis (L.) Merr. et Perry;
         syn: *Eugenia malaccensis* L.

FT 309   **Rose apple**
            Syzigium jambos (L.) Alston;
            syn: *Eugenia jambos* L.

FT 310   **Sea grape**
            Coccoloba uvifera Jacq.

FT 4121   **St. John's bread,** see Carob

FT 311   **Surinam cherry**
            Eugenia uniflora L.

FT 4123   **Tamarillo,** see Tree tomato

FT 4125   **Tree strawberry,** see Arbutus berry

FT 312   **Tree tomato**
            Cyphomandra betacea (Cav.) Sendt

Assorted tropical and sub-tropical fruits - inedible peel

<u>Class A</u>
Type 1   Fruits Group 6 Group Letter Code FI

The assorted tropical and sub-tropical fruits - inedible peel are derived from the immature or mature fruits of a large variety of perennial plants, usually shrubs or trees. Fruits are fully exposed to pesticides applied during the growing season (period of fruit development) but the edible portion is protected by skin, peel or husk. The edible part of the fruits may be consumed in a fresh or processed form.

<u>Portion of the commodity to which the MRL applies (and which is analyzed)</u>: *Whole fruit unless qualified: e.g., banana pulp. Pineapple after removal of crown. Avocado, mangos and similar fruit with hard seeds: Whole commodity after removal of stone but calculated on whole fruit.*

Group 6   Assorted tropical and sub-tropical fruits - inedible peel

<u>Code No.</u>   <u>Commodity</u>

FI   30   **Assorted tropical and sub-tropical fruits-inedible peel**

FI 5298   **Achiote,** see Annatto

FI   325   **Akee apple**
                 Blighia sapida Koenig

FI   326   **Avocado**
                 Persea americana Mill.

FI   327   **Banana**
                 Subsp. and cultivars of *Musa* spp. and hybrids

FI   328   **Banana, Dwarf**
                 Musa hybrids, AAA group;
                 syn: M. *cavendishii* Lambert: M. *nana* Lour.

FI   329   **Breadfruit**
                 Artocarpus communis J.R. et G. Forster:
                 syn: *Artocarpus altilis* (Parkinson) Fosberg

FI   330   **Canistel**
                 Pouteria campechiana (HBK.) Baenhi; this species includes former *Lacuma nervosa* A. DC. and L. *salicifolia* HBK.

FI 331   **Cherimoya**
              Annona cherimola Mill.

FI 4127   **Chinese gooseberry**, see Kiwifruit

FI 4128   **Chinese persimmon**, see subgroup 5 Persimmon, Japanese

FI 332   **Custard apple**
              Annona reticulata L.

FI 333   **Doum or Dum palm**
              Hyphaene thebaica (L.) Mart.

FI 334   **Durian**
              Durio zibethinus Murr.

FI 4129   **Egg fruit**, see Canistel

FI 371   **Elephant apple**
              Feronia limonia (L.) Swing
              syn: *Feronia elephantum*: *Limonia acidissima*

FI 335   **Feijoa**
              Feijoa sellowiana (O. Berg) O. Berg;
              syn: *Acca sellowiana* (D. Berg) Burret

FI 4131   **Genip**, see Marmaladebox

FI 4132   **Granddilla**, see Passion fruit

FI 4134   **Guanabana**, see Soursop

FI 336   **Guava**
              Psidium guajava L.

FI 337   **Ilama**
              Annona diversifolia Saff.

FI 4133   **Indian Fig**, see Prickly pear

FI 4136   **Indian wood apple**, see Elephant apple

FI 338   **Jackfruit**
             Artocarpus heterophyllus Lam.;
             syn: A. *integra* (Thunb.) Merr.: A *integrifolia* L.f.

FI 339   **Jambolan**
             Zyzigium cumini (L.) Skeels
             syn: *Eugenia cuminii* (L.) Druce; *Eugenia obtusifolia* Roxb.

FI 340   **Java apple**
             Eugenia javanica Lam;
             syn: *Syzigium samarangense* (BL.) Merr. & Perry

FI 341   **Kiwifruit**
             Actinidia deliciosa (A. Chev.) Liang et Ferguson;
             syn: A. *chinensis* Planck

FI 342   **Longan**
             Nephelium longana (Lam.) Camb.;
             syn: *Euphoria longana* Lam.

FI 343   **Litchi**
             Litchi chinensis Sonn.
             syn: *Nephelium litchi* Camb.

FI 4135   **Lulo,** see Naranjilla

FI 4138   **Malay apple,** see Pomarac

FI 344   **Mammey apple**
             Mammea americana L.

FI 345   **Mango**
             Mangifera indica L.

FI 346   **Mangostan**
             Garcinia mangostana L.

FI 4137   **Mangosteen,** see Mangostan

FI 347   **Marmaladedos**
             Genipa americana L.

FI 348   **Mombin, Yellow**
                 Spondias lutea L.;
                 syn: S. *mombin* L.

FI 349   **Naranjilla**
                 Solanum quitoense Lam.

FI 4139   **Papaw or Pawpaw**, see Papaya

FI 350   **Papaya**
                 Carica papaya L.

FI 351   **Passion fruit**
                 Cultivars of *Passiflora edulis* Sims

FI 352   **Persimmon, American**
                 Diospyros virginiana L.

FI 4141   **Persimmon, Japanese**, see Group 5

FI 353   **Pineapple**
                 Ananas comosus (L.) Merril;
                 syn: A *sativus* (L.) Lindl.

FI 4143   **Pineapple guava**, see Feijoa

FI 354   **Plantain**
                 Musa X paradisiaca L., *var. sapientum* (L.) Kuntze

FI 355   **Pomegranate**
                 Punica granatum L.

FI 356   **Prickly pear**
                 Opuntia ficus-indica (L.) P. Miller

FI 357   **Pulasan**
                 Nephelium mutabile Bl.

FI 4145   **Quito orange**, see Naranjilla

FI 358   **Rambutan**
                 Nephelium lappaceum L.

FI 359   **Sapodilla**
Manilkara achras (Mill.) Fosberg:
syn: *Achras zapota* L.

FI 360   **Sapote, Black**
Diospyros ebenaster Retz.

FI 361   **Sapote, Green**
Calocarpum viride Pitt.

FI 362   **Sapote, Mammey**
Calocarpum sapota (Jacq.) Merr.

FI 363   **Sapote, White**
Casimiroa edulis Llave & Lex

FI 364   **Sentul**
Sandoricum koetjape Merr.:
syn: S. *indicum* Car.

FI 4147 **Sesso vegetal**, see Akee apple

FI 365   **Soursop**
Annona muricata L.

FI 366   **Spanish lime**
Melicoccus bijugatus Jacq.;
syn: *Melicocca bijuga* L.

FI 367   **Star apple**
Chrysophyllum cainito L.

FI 4149 **Strawberry peach**, see Kiwifruit

FI 368   **Sugar apple**
Annonn squamosa L.

FI 4151 **Sweetsop**, see Sugar apple

FI 369   **Tamarind**
Tamarindus indica L.

FI 370   **Tonka bean**
Dipteryx odorata (Aubl.) Willd.: D. *oppositifolia* (Aubl.) Willd

TYPE 2   VEGETABLES

Vegetables are foods derived from many different kinds of plants mostly annual and usually cultivated, commonly known by custom and tradition as "vegetables".

In several countries, some of these commodities grown on large areas are distinguished as "field crops" or arable crops e.g. sugar beet. For the sake of convenience in this guide such crops are classified under Type 2 Vegetables.

Exposure to pesticides is dependent on the particular part of the plant used for food and the growing practices.

Vegetables may be consumed in whole or in part and in the form of fresh, dried or processed foods.

Bulb vegetables

Class A
Type 2   Vegetables Group 9 Group Letter Code VA

Bulb vegetables are pungent highly flavoured foods derived from fleshy scale bulbs in some commodities including stem and leaves), of the *genus Allium* of the lily family (*Liliaceae*). Bulb fennel is included in this group; the bulb-like growth of this commodity gives rise to similar residues.

The subterranean parts of the bulbs and shoots are protected from direct exposure to pesticides during the growing season.

The entire bulb may be consumed after removal of the parchment-like skin. The leaves and stems of some species or cultivars may also be consumed.

Portion of the commodity to which the MRL applies (and which is analyzed): *Bulb/dry onions and garlic: Whole commodity after removal of roots and adhering soil and whatever parchment skin is easily detached. Leeks and spring onions: Whole vegetable after removal of roots and adhering soil.*

Group 9   Bulb vegetables

Code No.   Commodity

VA 35   **Bulb vegetables**

VA 36   **Bulb vegetables, except Fennel, Bulb**

VA 4153   **Carosella**, see Fennel, Italian

VA 4155   **Chives**, see Group 027: Herbs

VA 4157   **Chives, Chinese**, see Group 027: Herbs

VA 380   **Fennel, Bulb**
Foeniculum vulgare Mill.;
syn: F. *officinale* All.
- var. *dulce* (Mill.) Thell.: syn: F. *dulce* Mill.;
- var. *azoricum* (Mill.) Thell.; syn: F. *azoricum* Mill.

VA 4159   **Fennel, Italian**, see Fennel, bulb
Foeniculum vulgare Mill., var. *azoricum* (Mill.) Thell.;
syn: F. *azoricum* Mill.

VA 4161   **Fennel, Roman**, see Fennel, bulb
Foeniculum vulgare Mill., *var. dulce* (Mill.) Thell.:
syn: F. *dulce* Mill.

VA 4163   **Fennel, Sweet**, see Fennel, Roman

VA 381   **Garlic**
Allium sativum L.

VA 382   **Garlic, Great-headed**
Allium ampeloprasum L., var. *ampeloprasum*

VA 4165   **Japanese bunching onion**, see Welsh onion

VA 383   **Kurrat**
Allium ampeloprasum L., var. kurrat Schweinf. ex Krause

VA 384   **Leek**
Allium porrum L.;
syn: A. *ampeloprasum* L., var. *porrum* (L.) Gay

VA 4167   **Multiplying onion**, see Onion, Welsh

VA 385   **Onion, Bulb**
Allium cepa L. var. *cepa*, various cultivars

VA 386   **Onion, Chinese**
            Allium chinense Don.:
            syn: A. bakeri Regel

VA 4169   **Onion, Egyptian**, see Tree onion

VA 387   **Onion, Welsh**
            Allium fistulosum L.

VA 4171   **Rakkyo**, see Onion, Chinese

VA 388   **Shallot**
            Allium ascalonicum L.;
            syn: A. cepa L., var. *aggregatum* Don.

VA 389   **Spring onion**
            Allium cepa L., various cultivars, a.o. White Lisbon; White Portugal

VA 390   **Silverskin onion**
            Allium cepa L., var.

VA 391   **Tree onion**
            Allium cepa L., var. *proliferum* Targioni-Tozzetti;
            syn: A. *cepa* L., var. *bulbiferum* Bailey;
            A. *cepa* L., var. *viviparum* (Metz.) Alef

Brassica (cole or cabbage) vegetables,
Head cabbages, Flowerhead brassicas

Class A
Type 2   Vegetables   Group 010   Group Letter Code VB

Brassica (cole or cabbage) vegetables and flowerhead brassicas are foods derived from the leafy heads, stems and immature inflorescences of plants belonging to the *genus Brassica* or the family *Cruciferae*. Although Kohlrabi does not comply fully with the description above, for convenience and because of the similarity in residue behaviour the commodity is classified in this group. Kohlrabi is a tuber-like enlargement of the stem.

The edible part of the crop is partly protected from pesticides applied during the growing season by outer leaves, or skin (Kohlrabi).

The entire vegetable after discarding obviously decomposed or withered leaves may be consumed.

Portion of the commodity to which the MRL applies (and which is analyzed): *Head cabbages and Kohlrabi: Whole commodity is marketed, after removal of obviously decomposed or withered leaves. Cauliflower and broccoli: flower heads (immature inflorescence only). Brussels sprouts: "buttons" only.*

Group 010   Brassica (cole or cabbage) vegetables,
**   Head cabbages, Flowerhead brassicas**

Code No.   Commodity

VB 40   **Brassica (cole or cabbage) vegetables, Head cabbages, Flowerhead brassicas**

VB 41   **Cabbages, Head**
                  Brassica oleracea L., *convar. capitata* L., several var. and cvs.

VB 42   **Flowerhead brassicas** (includes Broccoli: Broccoli, Chinese and Cauliflower)

VB 400   **Broccoli**
                  Brassica oleracea L., convar *botrytis* L.,
                  var. *italica* Plenck

VB 401   **Broccoli, Chinese**
                  Brassica campestris L., var. *alboglabra* Bailey

VB 4173   **Broccoli, Sprouting**, see Broccoli

VB 402   **Brussels sprouts**
             Brassica oleracea L., convar. *oleracea* L.,
             var. *gemmifera* DC.

VB 4175   **Cabbage,** see Cabbages, Head

VB 4177   **Cabbage, Green,** see Cabbage, Savoy

VB 4179   **Cabbage, Red,** see Cabbages, Head
             Brassica oleracea L.; convar. *capitata* L., var. *rubra*

VB 4181   **Cabbage, Oxhead,** see Cabbages Head
             Brassica oleracea L., convar. *capitata* L, var. *alba, forma conica*

VB 4183   **Cabbage, Pointed,** see Cabbage, Oxhead

VB 4185   **Cabbage, White,** see Cabbages, Head
             Brassica oleracea L., convar. *capitata* L., var. *alba*

VB 403   **Cabbage, Savoy,** see also Cabbages, Head
             Brassica oleracea L., convar. *capitata* L., var. *sabauda* L.

VB 4187   **Cabbage, Yellow,** see Cabbage, Savoy

VB 404   **Cauliflower,** see also Flowerhead brassicas
             Brassica oleracea L., convar. *botrytis* L., var. *botrytis* L., several cvs (white
             and green)

VB 4189   **Cauliflower, Green,** see Cauliflower

VB 4191   **Kailan,** see Broccoli, Chinese

VB 405   **Kohlrabi**
             Brassica oleracea L., convar. *acephala*, var. *gongylodes*

Fruiting vegetables, Cucurbits

<u>Class A</u>
Type 2   Vegetables Group 011 Group Letter VC

Group 011. Fruiting vegetables, Cucurbits are derived from the immature or mature fruits of various plants, belonging to the botanical family *Cucurbitaceae*; usually these are annual vines or bushes.

These vegetables are fully exposed to pesticides during the period of fruit development.

The edible portion of those fruits of which the inedible peel is discarded before consumption is protected from most pesticides, by the skin or peel, except from pesticides with a systemic action.

The entire fruiting vegetable or the edible portion after discarding the inedible peel may be consumed in the fresh form or after processing. The entire immature fruit of some of the fruiting vegetables species may be consumed, whereas only the edible portion of the mature fruit of the same species, after discarding the then inedible peel, is consumed.

<u>Portion of the commodity to which the MRL applies (and which is analyzed)</u>: *Whole commodity after removal of stems.*

Group 011   Fruiting vegetables, Cucurbits

<u>Code No.</u>   <u>Commodity</u>

VC 45   **Fruiting vegetables, Cucurbits**

VC 46   **Melons, except Watermelon**
              Several var. and cultivars of *Cucumis melo* L.

VC 420   **Balsam apple**
              Momordica balsamina L.

VC 421   **Balsam pear**
              Momordica charantia L.

VC 4193   **Bitter cucumber,** see Balsam pear

VC 4195   **Bitter gourd,** see Balsam pear

VC 4197   **Bitter melon,** see Balsam pear

VC 422   **Bottle gourd**
Lagenaria siceraria (Molina) Standl.:
syn: L. *vulgaris* Ser.; L. *leucantha* (Duch.) Rusby

VC 4199   **Cantaloupe,** see Melons
Cucumis melo L., var. *cantaloupensis* Naud.

VC 4201   **Casaba or Casaba melon,** see Subgroup Melons, except Watermelon
Cucumis melo L., var. *inodorus* Naud.

VC 423   **Chayote**
Sechium edule (Jacq.) Schwartz;
syn: *Chayota edulis* Jacq.

VC 4203   **Christophine,** see Chayote

VC 4205   **Citron melon,** see Watermelon
Citrullus lanatus (Thunb.) Mansf., var. *edulis*;
syn: *Citrullus edulis* Pang.

VC 4207   **Courgette,** see Squash, Summer

VC 424   **Cucumber**
Cucumis sativus L.; English and forcing cucumber cultivars

VC 4209   **Cucuzzi,** see Bottle Gourd

VC 4211   **Cushaws,** see Pumpkins
Mature cultivars of *Cucurbita mixta* Pang.

VC 425   **Gherkin**
Cucumis sativus L.; pickling cucumber cultivars

VC 426   **Gherkin, West Indian**
Cucumis anguria L.

VC 427   **Loofah, Angled**
Luffa acutangula (L.) Roxb.

VC 428   **Loofah, Smooth**
Luffa cylindrica (L.) M.J. Roem:
syn: L. *aegyptiaca* Mill.

VC 4213   **Marrow,** see Squash, Summer
            Cucurbita pepo, several cultivars

VC 4215   **Melon, Crenshaw,** see Subgroup Melons, except Watermelon
            Cultivar of *Cucumis melo* L., var. *reticulatus* Naud.

VC 4217   **Melon, Honey Ball**, see Subgroup Melons, except Watermelon
            Cultivar of *Cucumus melo* L., var. *reticulatus* Naud.

VC 4219   **Melon, Honeydew,** see Subgroup Melons, except Watermelon
            Cultivar of Winter, or White-skinned Melon
            Cucumis melo L., var. *inodorus* Naud.

VC 4221   **Melon, Mango,** see Subgroup Melons, except Watermelon
            Cucumis melo L., var. chito Naud.

VC 4223   **Melon, Netted,** see Subgroup Melons, except Watermelon
            synonym of Muskmelon, see there

VC 4225   **Melon, Oriental Pickling,** see Subgroup Melons, except Watermelon
            Cucumis melo L., var. *conomon* Mak.

VC 4227   **Melon, Persian,** see Subgroup Melons, except Watermelon
            Cultivar of *Cucumis melo* L., var. *reticulatus* Naud.

VC 4229   **Melon, Pomegranate,** see Subgroup Melons, except Watermelon
            Cucumis melo L., var. *dudaim* Naud.

VC 4231   **Melon, Serpent,** see Subgroup Melons, except Watermelon
            Cucumis melo L., var. *flexuosus* Naud.

VC 4233   **Melon, Snake,** see Subgroup Melons, except Watermelon
            synonym of Melon, Serpent

VC 4235   **Melon, White-skinned,** see Subgroup Melons, except Watermelon
            Cultivars of *Cucumis melo* L. var. *inodorus* Naud.

VC 4237   **Melon, Winter,** see Subgroup Melons, except Watermelon
            synonym of Melons, White skinned, see there

VC 4239   **Muskmelon,** see Subgroup Melons, except Watermelon
            Cultivar of *Cucumis melo* L. var. *reticulatus* Naud.

VC 4241   **Patisson,** see Squash, White Bush
           Cucurbita pepo L., var. *patissonina*

VC 429   **Pumpkins**
          Mature cultivars of *Cucurbita maxima* Duch. ex Lam.;
          C. *mixta* Pang.; C. *moschata* (Duch. ex Lam.) Duch. ex Poir.
          and C. *pepo* L.

VC 4243   **Sinkwa or Sinkwa towel gourd,** see Loofah, Angled

VC 430   **Snake gourd**
          Trichosanthes cucumerina L.;
          syn: T. *anguina* L.

VC 4245   **Sponge gourd,** see Loofah, Smooth

VC 4247   **Squash,** see Squash, Summer, and Winter Squash

VC 431   **Squash, Summer**
          Cucurbita pepo L., var. *melopepo* Alef
          several cultivars, immature

VC 4249   **Squash,** White bush, see Squash, Summer

VC 4251   **Vegetable spaghetti,** see Pumpkins
           Cucurbita pepo

VC 4253   **Vegetable sponge,** see Loofah, Smooth

VC 432   **Watermelon**
          Citrullus lanatus (Thunb.) Mansf.
          syn: C. *vulgaris* Schrad.; *Colocynthis citrullus* (L.)
          O. Ktze.

VC 4255   **Wax gourd**
           Benincasa hispida (Thunb.) Cogn.;
           syn: B. *cerifera* Savi

VC 4257   **West Indian gherkin,** see Gherkin, West Indian

VC 4259   **Winter melon,** see Melon, Winter

VC 433   **Winter squash,** see also Pumpkins
          Mature cultivars of *Cucurbita maxima* Duch. ex Lam.;
          C. *mixta* Pang.; C. *moschata* (Duch. ex Lam.) Duch. ex Poir. and C. *pepo* L.

VC 4261   **Zucchetti,** see Squash, Summer

VC 4263   **Zucchini,** see Squash, Summer

Fruiting vegetables, other than Cucurbits
(not including pods of leguminous vegetables, see Group 014)

<u>Class A</u>
Type 2   Vegetables Group 012 Group Letter Code V0

Group 012. Fruiting vegetables, other than Cucurbits are derived from the immature and mature fruits of various plants, usually annual vines or bushes. The group includes edible fungi and mushrooms, being comparable organs of lower plants. Many plants of this group belong to the botanical family *Solanaceae*.

This group does not include fruits of vegetables of the botanical family *Cucurbitaceae* or the pods of vegetables of the *Leguminosae*-family.

The vegetables of this group are fully exposed to pesticides applied during the period of fruit development, except those of which the edible portion is covered by husks, such as sweet corn, ground cherries (*Physalis* spp.). The latter fruiting vegetables are protected from most pesticides by the husk except from pesticides with a systemic action.

The entire fruiting vegetable or the edible portion after discarding husks or peels may be consumed in a fresh form or after processing.

<u>Portion of the commodity to which the MRL applies (and which is analyzed)</u>: *Whole commodity after removal of stems. Mushrooms: Whole commodity. Sweet corn and fresh corn: kernels plus cob without husk.*

Group 012   Fruiting vegetables, other than Cucurbits

<u>Code No.</u>   <u>Commodity</u>

VO 50   **Fruiting vegetables,** other than Cucurbits

VO 51   **Peppers**
             Subgroup including Peppers, Chili and Peppers, Sweet

VO 4265   **Alkekengi,** see Ground cherries
             Physalis alkekengi L.

VO 4267   **Aubergine,** see Egg plant

VO 4269   **Bell pepper,** see Peppers, Sweet

VO 4271   **Cape gooseberry,** see Ground cherries
             Physalis peruviana L.

VO 4273   **Cherry pepper,** see Peppers, Chili
           Capsicum annuum L., var. *acumimata* Fingerh.

VO 4275   **Cherry tomato,** see Ground cherries

VO 4277   **Chili peppers,** see Peppers, Chili

VO 4279   **Chinese lantern plant,** see Ground cherries

VO 4281   **Cluster pepper,** see Peppers, Chili
           Capsicum annuum L., var. *fasciculatum* (Sturt.) Irish

VO 4283   **Cone pepper,** see peppers, Chili

VO 4285   **Corn-on-the-cob,** see Sweet corn (corn-on-the-cob)

VO 440   **Egg plant**
           Solanum melongena L., var. *melongena* L.

VO 449   **Fungi, Edible** (not including mushrooms)
           According to Codex Stan. 35-1981: various edible species of fungi, mainly
           wild, among others *Boletus edulis*; other *Boletus* spp, *Morchella* spp, *Pleurotus
           ostreatus*

VO 4287   **Fungus "Chanterelle",** see Fungi, Edible
           Cantharellus cibarius (Codex Stan. 40-1981)

VO 4289   **Golden berry,** see Ground cherries
           Physalis peruviana L.

VO 441   **Ground cherries**
           Physalis alkekengi L.; Ph, *ixocarpa* Brot. ex Horn, Ph. *peruviana* L.

VO 4291   **Husk tomato,** see Ground cherries

VO 4293   **Lady's finger,** see Okra

VO 4295   **Melon pear,** see Pepino

VO 450   **Mushrooms**
           Cultivated cultivars of *Agaricus* spp,
           syn: *Psalliota* spp., mainly *Agaricus bisporus* (definition Codex Stan. 55-1981)

VO 4297   **Naranjilla,** see Group 6 Assorted tropical and sub-tropical fruits - inedible peel

VO 442   **Okra**
          Hibiscus esculentus L.;
          syn: *Abelmoschus esculentus* (L.) Moench.

VO 4299   **Paprika,** see Peppers, Sweet

VO 443   **Pepino**
          Solanum muricatum L.

VO 444   **Peppers, Chili**
          Capsicum annuum L.; several pungent cultivars

VO 4301   **Peppers, Long,** see Peppers, Sweet
          Capsicum annuum L., var. *longum* (D.C.) Sendt.

VO 445   **Peppers, Sweet** (including pimento or pimiento)
          Capsicum annuum, var. *grossum* (L.) Sendt. and var. *longum* (D.C.) Sendt.

VO 4303   **Pimento or Pimiento,** see Peppers, Sweet

VO 4305   **Quito Orange,** see Naranjilla

VO 446   **Roselle**
          Hibiscus sabdariffa L., var. *sabdariffa* L.

VO 4307   **Strawberry tomato,** see Ground cherries

VO 447   **Sweet corn (corn-on-the-cob),** see definition in Codex Stan. 133-1981
          Zea mays L., var. *saccharata* Sturt;
          syn: *Zea mays* L., var *rugosa* Bonof.

VO 1275   **Sweet corn (kernels),** see definition in Codex Stan. 132-1981

VO 4309   **Tomatillo, see Ground cherries**
          Physalis ixocarpa Brot. ex Horn.

VO 448   **Tomato**
          Lycopersicon esculentum Mill.;
          syn: *Solanum lycopersicum* L.

VO 4311   **Tree melon,** see Pepino

Leafy vegetables (including Brassica leafy vegetables)

<u>Class A</u>
Type 2   Vegetables Group 013 Group Letter Code VL

Group 013. Leafy vegetables are foods derived from the leaves of a wide variety of edible plants, usually annuals or biennials. They are characterized by a high surface:weight ratio. The leaves are fully exposed to pesticides applied during the growing season.

The entire leaf may be consumed, either fresh or after processing or household cooking.

<u>Portion of the commodity to which the MRL applies (and which is analyzed)</u>: *Whole commodity as usually marketed, after removal of obviously decomposed or withered leaves.*

Group 013   Leafy vegetables (including Brassica leafy vegetables)

<u>Code No.</u>   <u>Commodity</u>

VL 53   **Leafy vegetables**

VL 54   **Brassica leafy vegetables**
              Brassica spp.

VL 460   **Amaranth**
              among others *Amaranthus dubius* Mart. ex Thell.:
              A. *cruentus* L.; A. *tricolor* L., several var.

VL 4313   **Amsoi,** see Indian Mustard

VL 4315   **Arrugula,** see Rucola

VL 421   **Balsam pear leaves**
              Momordica charantia

VL 4317   **Beet leaves,** see Chard

VL 461   **Betel leaves**
              Piper betle L.

VL 4319   **Bitter cucumber leaves,** see Balsam pear leaves

VL 4321   **Blackjack**
              Bidens pilosa L.

VL 4323   **Bledo,** see Amaranth

VL 4325   **Borecole,** see Kale, curly

VL 462   **Box thorn**
           Lycium chinense

VL 4327   **Broccoli raab**
           Similar to Turnip greens, see there
           Brassica campestris L., *ruvo* group

VL 463   **Cassava leaves**
           Manihot esculenta Crantz

VL 4329   **Celery cabbage,** see Chinese cabbage

VL 4331   **Celery mustard,** see Pak-choi

VL 464   **Chard**
           Beta vulgaris L , var. *vulgaris*;
           syn: B. *vulgaris* L., var. *cicla* L.

VL 465   **Chervil**
           Anthriscus cerefolium (L.) Hoffmann

VL 469   **Chicory leaves** (green and red cultivars
           Cichorium intybus L., var. *foliosum* Hegi

VL 467   **Chinese cabbage,** (type Pe-tsai)
           Brassica pekinensis (Lour.) Ruprecht
           syn: B. *campestris* L., ssp *pekinensis* (Lour.) Olson

VL 468   **Choisum**
           Brassica campestris L., var. *parachinensis* (Bailey) Sinsk.

VL 4332   **Collard,** see Kale

VL 470   **Corn salad**
           Valerianella locusta (L.) *Laterrade*;
           syn: V. *olitoria*

VL 510   **Cos lettuce**
           Lactuca sativa L.: var. *romana*
           L. *sativa* L.; var. *longifolia*

VL 4333   **Cowslip,** see Marsh marigold
             Caltha palustris L.

VL 472   **Cress, Garden**
             Lepidium sativum

VL 4335   **Crisphead lettuce,** see Lettuce, Head

VL 4337   **Curly kale,** see Kale, curly

VL 4339   **Cutting lettuce,** see Lettuce, leaf

VL 474   **Dandelion**
             Taraxacum officinale Weber

VL 475   **Dock**
             Rumex spp.; also *Rumex* hybrids

VL 476   **Endive**
             Cichorium endivia L.

VL 4341   **Endive, broad or plain leaved,** see Endive
             Cichorium endivia L., var. *latifolium* Lamarck

VL 4343   **Endive, curled,** see Endive
             Cichorium endivia L., var. *crispum* Lamarck

VL 4345   **Fennel,** see Group 027 Herbs

VL 4347   **Fennel, Bulb,** see Group 9 Bulb vegetables

VL 4349   **Garden cress,** see Cress, Garden

VL 477   **Goosefoot**
             Chenopodium spp.

VL 4351   **Gow Kee,** see Box thorn

VL 269   **Grape leaves**
             Vitis vinifera L.

VL 478   **Indian mustard**
             Brassica juncea (L.) Czern. & Coss.

VL 4353   **Jamaican sorrel**, see Roselle leaves

VL 479   **Japanese greens,** various species, a.o.
          Chrysanthemum coronarium L.; Turnip greens (see there)
          Mizuma, Indian Mustard and Komatsuna

VL 480   **Kale** (including among others: Collards, Curly kale, Scotch kale, thousand-headed
          kale; not including Marrow-stem kale, no. AV 1052 Miscellaneous Fodder and
          forage crops
          Brassica oleracea L., convar. *acephala* (D.C.) Alef., var. *acephala*

VL 4355   **Kale, curly,** see Kale
           Brassica oleracea L., convar. *acephala* (D.C.) Alef., var. *sabellica* L.

VL 507   **Kangkung**
          Ipomoea aquatica Forsk.;
          syn: I. *reptans* Poir.

VL 481   **Komatsuma**
          Brassica pervirides H.L. Bail.

VL 4357   **Lambs lettuce,** see Corn salad

VL 482   **Lettuce, Head**
          Lactuca sativa L., var. *capitata*

VL 483   **Lettuce, Leaf**
          Lactuca sativa L., var. *crispa* L.;
          syn: L. *sativa*, var. *foliosa*

VL 4359   **Lettuce, Red,** see Lettuce, Head
           Red cultivar of *Lactuca sativa*, var. *capitata*

VL 484   **Mallow**
          Malva verticillata L .; syn : M. *crispa* L .;
          M. *mohileviensis* Graebn., M. *pamiroalaica* Ilj. and M. *sylvestris* L.

VL 471   **Marsh marigold**
          Caltha palustris L.

VL 4361   **Matrimony vine,** see Box thorn

VL 485   **Mustard greens**
among others
Brassica juncea (L.) Czern & Coss ssp. *juncea*

VL 4363   **Mustard, Indian,** see Indian Mustard

VL 4364   **Mustard spinach,** see Komatsuma

VL 486   **New Zealand spinach**
Tetragonia tetragonioides (Pallas) O. Kuntze;
syn: T. *expansa* Murr.

VL 487   **Nightshade, Black**
Solanum nigrum L.

VL 4365   **Namenia,** see Turnip greens

VL 488   **Orach**
Atriplex hortensis L.

VL 466   **Pak-choi or Paksoi**
Brassica sinensis L.;
syn: B. *campestris*, ssp *chinensis* (L.) Makino

VL 4367   **Pak-tsai,** see Chinese cabbage, (type Pe-tsai)

VL 4368   **Pak-tsoi or Pak-soi,** see Pak-choi or Paksoi

VL 337   **Papaya leaves**
Carica papaya L.

VL 489   **Pepper leaves**
Piper umbellatum L. (Asia); P. *auritum* H.b.& K.;
P. *sanctum* (Miq.) Schlecht., both Centr. and S. America

VL 490   **Plantain leaves**
Plantago major L.

VL 4369   **Poke-berry leaves,** see Pokeweed

VL 491   **Pokeweed**
Phytolacca americana L.;
syn: P. *decandra* L.; P. *rivinoides* H. et B.

VL 492   **Purslane**
          Portulaca oleracea L., ssp. *sativa* (Haw) Celak.

VL 493   **Purslane, Winter**
          Claytonia perfoliata Donn ex Willd.;
          syn: *Montia perfoliata* Howell

VL 494   **Radish leaves** (including Radish tops)
          Raphanus sativus L., several varieties

VL 495   **Rape greens**
          Brassica napus L.

VL 4371   **Red leaved chicory,** see Chicory leaves

VL 4372   **Rocket salad,** see Rucola

VL 4374   **Roquette,** see Rucola

VL 446   **Roselle leaves**
          Hibiscus sabdariffa L.

VL 496   **Rucola**
          Eruca vesicaria (L.) Cav. ssp *sativa* mill.
          E. *sativa* L.

VL 497   **Rutabaga greens**
          Brassica napobrassica (L) Mill.;
          syn: B. *napus* L., var. *napobrassica* (L.) Rchb.

VL 498   **Salsify leaves**
          Tragopogon porrifolium L.

VL 499   **Sea kale**
          Crambe maritima L.

VL 500   **Senna leaves**
          Cassia senna L.;
          syn: C. *acutifolia* Del.

VL 4373   **Silver beet,** see Chard
          Beta vulgaris L.; var. *flaveseens*

VL 4388   **Sorrel, Jamaican,** see Roselle leaves

VL 501   **Sowthistle**
           Sonchus oleraceus L.

VL 502   **Spinach**
           Spinacia oleracea L.

VL 4375   **Spinach beet,** see Chard
           beta vulgaris L.; var. *vulgaris*

VL 503   **Spinach, Indian**
           Basella alba L.;
           syn: B. *rubra* L.

VL 4377   **Sugar loaf,** see Chicory leaves

VL 508   **Sweet potato,** leaves
           Ipomoea batatas (L.) Poir.

VL 4379   **Swiss chard,** see Chard

VL 504   **Tannia leaves**
           Xanthosoma sagittifolium (L.) Schott;
           syn: X. edule (Mey) Schott X. xanthorrhizon (Jacq.)
           C. Koch; *Arum sagittaefolium* L.

VL 505   **Taro leaves**
           Colocasia esculenta (L.) Schott

VL 4381   **Tendergreen,** see Turnip greens

VL 4383   **Tsai shim,** see Choisum

VL 4385   **Tsoi sum,** see Choisum

VL 506   **Turnip greens**
           Brassica rapa L., var. *rapa*;
           syn: B. *campestris* L., var. *rapifera* Metz.

VL 4387   **Vine spinach,** see Spinach, Indian

VL 473   **Watercress**
           Nasturtium officinale R. Br. and a hybrid of N. *officinalis* R. Br. x N.
           microphyllum (Boenningh.) Rchb.

VL 4389   **Water Spinach,** see Kangkung

VL 4391   **Yautia leaves,** see Tannia leaves

Legume vegetables

<u>Class A</u>
Type 2   Vegetables Group 014 Group Letter Code VP

Group 014. Legume vegetables are derived from the succulent seed and immature pods of leguminous plants commonly known as beans and peas.

Pods are fully exposed to pesticides during the growing season, whereas the succulent seed is protected within the pod from most pesticides, except pesticides with systemic action.

The succulent forms may be consumed as whole pods or as the shelled product.

<u>Portion of the commodity to which the MRL applies (and which is analyzed)</u>: *Whole commodity, unless otherwise specified.*

Group 014   Legume vegetables

<u>Code No.</u>   <u>Commodity</u>

VP 60   **Legume vegetables**

VP 61   **Beans, except broad bean and soya bean**
                   (green pods and immature seeds)
                   Phaseolus spp.

VP 62   **Beans, shelled**
                   (succulent = immature seeds)

VP 63   **Peas** (pods and succulent = immature seeds)
                   Pisum spp.; *Vigna* spp.

VP 64   **Peas, shelled** (succulent seeds)
                   Pisum spp.; *Vigna* spp.

VP 4393   **Angola pea (immature seed)**, see Pigeon pea·

VP 4395   **Asparagus bean (pods)**, see Yard-long bean

VP 4397   **Asparagus pea (pods)**, see Goa bean

VP 520   **Bambara groundnut** (immature seeds)
                   Voandzeia *subterranea* (L.) Thou.

VP 521   **Black gram (green pods)**
                Phaseolus mungo L.;
                syn: *Vigna mungo* (L.) Hepper

VP 4399   **Bonavist bean (young pods and immature seeds),**
                see Hyacinth bean

VP 522   **Broad bean (green pods and immature seeds)**
                Vicia faba L. subsp. *eu-faba*, var. major Harz and
                var. minor Beck.

VP 523   **Broad bean, shelled (succulent) (= immature seeds)**

VP 4401   **Butter bean (immature pods),** see Lima bean

VP 4401   **Cajan pea (young green seeds),** see Pigeon pea

VP 4401   **Catjang cowpea (immature pods and green seeds),** see Cowpea
                Vigna unguiculata (L.) Walp.;
                syn: *Dolichos* catjang Burm.; D. *unguiculatus* L.

VP 524   **Chick-pea (green pods)**
                Cicer arietinum L.

VP 525   **Cluster bean (young pods)**
                Cyamopsis tetragonoloba (L.) Taub;
                syn: C. *psoralioides* DC.

VP 526   **Common bean (pods and/or immature seeds)**
                Phaseolus vulgaris L., several cultivars

VP 527   **Cowpea (immature pods)**
                Vigna unguiculata L., Cv-group *unguiculata*

VP 4403   **Dwarf bean (immature pods and/or seeds),** see Common bean

VP 4405   **Edible-podded pea,** see Podded pea

VP 4407   **Fava bean (green pods and immature beans),** see Broad bean

VP 4409   **Field bean (green pods),** see Common bean

VP 4411   **Flageolet (fresh beans),** see Common bean

VP 4413   **Four-angled bean (immature pods),** see Goa bean

VP 4415   **French bean (immature pods and seeds),** see Common bean

VP 4417   **Garbanzos,** see Chick-pea

VP 528   **Garden pea (young pods)** (= succulent, immature seeds)
                 Pisum sativum L., subsp. *hortense* (Neilr.) A. et G.;
                 syn: P. *sativum* L., subsp. *sativum* L.

VP 529   **Garden pea, shelled** (succulent seeds)
                 for scientific name see above

VP 530   **Goa bean (immature pods)**
                 Psophocarpus tetragonolobus (L.) DC.

VP 4419   **Gram (green pods),** see Chick-pea

VP 4421   **Green bean (green pods and immature seeds),** see Common bean

VP 4423   **Green gram (green pods),** see Mung bean

VP 4425   **Guar (young pods),** see Cluster bean

VP 4427   **Haricot bean (green pods, and/or immature seeds),**
                 see Common bean

VP 4429   **Horse bean (green pods and/or immature seeds),**
                 see Broad bean

VP 531   **Hyacinth bean (young pods, immature seeds)**
                 Dolichos lablab L.;
                 syn: *Lablab niger* Medik: L. *vulgaris Savi*

VP 532   **Jack bean (young pods, immature seeds)**
                 Canavalia ensiformis (L.) DC.

VP 4431   **Kidney bean (pods and/or immature seeds),** see Common bean

VP 4433   **Lablab (young pods; immature seeds),** see Hyacinth bean

VP 533   **Lentil (young pods)**
                 Lens esculenta Moench.;
                 syn: L. *culinaris* Medik; *Ervum lens* L.

VP 534   **Lima bean (young pods and/or immature beans)**
            Phaseolus lunatus L.;
            syn: Ph. *limensis* Macf.; Ph. *inamoenus* L.

VP 545   **Lupin**
            Lupinus ssp, sweet spp., varieties and cultivars with a low alkaloid content

VP 4435   **Mangetout or Mangetout pea,** see Podded pea

VP 4435   **Manila bean (immature pods),** see Goa bean

VP 535   **Mat bean (green pods, mature, fresh seeds)**
            Phaseolus aconitifolius Jacq.;
            syn: Ph. *trilobus* Ait; Vigna *aconitifolius* (Jacq.) Verde.

VP 4437   **Moth bean, see Mat bean**

VP 536   **Mung bean (green pods)**
            Phaseolus aureus Roxb;
            syn: *Vigna radiata* (L.) Wilczek, var. *radiata*;
            V. *aureus* (Roxb.) Hepper

VP 4439   **Navy bean (young pods and/or immature seeds),**
            see Common bean

VP 4441   **Pea,** see Garden pea

VP 4443   **Pigeon bean (green pods and immature seeds),** see Broad bean
            Vicia faba L. subsp. *eu-faba*, var. minor Beck

VP 537   **Pigeon pea (green pods and/or young green seeds)**
            Cajanus cajan (L.) Millsp.;
            syn: C. *indicus* Spreng

VP 538   **Podded pea (young pods)**
            Pisum sativum L., subsp. *sativum* var. *axiphium;*
            P. *sativum* L., subsp. *sativum*, var. *sacharatum*

VP 4447   **Red gram (green pods and/or young green seeds),** see Pigeon pea

VP 539   **Rice bean** (young pods)
            Vigna umbellata (Thunb.) Ohwi et Ohashi:
            syn: V. *calcarata* (Roxb.) Kurz *Phaseolus calcaratus* Roxb.

VP 4449   **Runner bean (green pods and seeds),** see Common bean

VP 540   **Scarlet runner bean (pods and seeds)**
             Phaseolus coccineus L.;
             syn: Ph. *multiflorus* Willd.

VP 4451   **Sieva bean (young pods and/or green fresh beans),** see Lima bean

VP 4453   **Snap bean (young pods),** see Common bean

VP 541   **Soya bean (immature seeds)**
             Glycine max (L.) Merr.;
             syn: G. *soja* Sieb. & Succ.: G. *hispida* (Moench) *Maxim.*:
             Soja max (L.) Piper

VP 4455   **Soybean,** see Soya bean

VP 4457   **Sugar pea (young pods),** see Podded pea
             Pisum sativum L., subsp. *sativum*, var. *sacharatum*

VP 542   **Sword bean (young pods and beans)**
             Canavalia gladiata (Jacq.) DC.

VP 4459   **Urd bean (green pods),** see Black gram

VP 4461   **Winged bean (immature pods),** see Goa bean

VP 543   **Winged pea (young pods)**
             Tetragonolobus purpureus Moench;
             syn: *Lotus tetragonolobus* L.

VP 4463   **Wrinkled pea,** see Garden pea
             Pisum sativum L., convar. *medullare*

VP 544   **Yard-long bean (pods)**
             Vigna unguiculata (L.) Walp, Cv-group *sesquipedalis*

Pulses

<u>Class A</u>
Type 2   **Vegetables Group 015 Group Letter Code VD**

Group 015.   Pulses are derived from the mature seeds, naturally or artificially dried, of leguminous plants known as beans (dry) and peas (dry).

The seeds in the pods are protected from most pesticides applied during the growing season except pesticides which show a systemic action.   The dried beans and peas however are often exposed to post-harvest treatments.

The dry pulses are consumed after processing or household cooking.

<u>Portion of the commodity to which the MRL applies (and which is analyzed)</u>: *Whole commodity*.

Group 015   Pulses

<u>Code No.</u>   <u>Commodity</u>

VD   70   **Pulses**

VD   71   **Beans (dry)**
            Phaseolus spp.; several species and cultivars

VD   72   **Peas (dry)**
            Pisum spp.; *Vigna* spp.

VD 560   **Adzuki bean (dry)**
            Phaseolus angularis (Willd.) Wight;
            syn: *Vigna angularis* (Willd.) Ohwi & Ohashi

VD 4465   **Angola pea**, see Pigeon pea

VD 520   **Bambara groundnut** (dry seed)
            Vigna subterranea (L.) Verde.;
            syn: *Voandzeia subterranea* (L.) Thou.

VD 4467   **Black-eyed pea**, see Cowpea

VD 521   **Black gram**
            Phaseolus mungo L.;
            syn: *Vigna mungo* (L.) Hepper

VD 4469   **Bonavist bean,** see Hyacinth bean

VD 523   **Broad bean (dry)**
           Vicia faba L., subsp. *eu-faba*, var. *major* Harz. and var. *minor* Beck

VD 4470   **Butter bean,** see Lima bean

VD 4471   **Cajan pea,** see Pigeon pea

VD 524   **Chick-pea (dry)**
           Cicer arietinum L.

VD 526   **Common bean (dry)**
           Phaseolus vulgaris L.

VD 527   **Cowpea (dry)**
           Vigna unguiculata (L.) Walp;
           syn: *V. sinensis* (L.) Savi ex Hassk.: *Dolichos sinensis* L.

VD 4473   **Dwarf bean (dry),** see Common bean (dry)

VD 4475   **Fava bean (dry),** see Broad bean (dry)

VD 4477   **Field bean (dry),** see Common bean (dry)

VD 561   **Field pea (dry)**
           Pisum sativum L., subsp. *arvense* (L.) A. et G.;
           syn: *Pisum arvense* L.

VD 4479   **Flageolet (dry),** see Common bean (dry)

VD 4481   **French bean,** see Group 06: Legume vegetables

VD 4483   **Geocarpa groundnut or Geocarpa bean,** see Kersting's groundnut

VD 4485   **Garden pea,** see Group 06: Legume vegetables

VD 4487   **Goa bean,** see Group 06: Legume vegetables

VD 4489   **Gram (dry),** see Chick-pea (dry)

VD 4491   **Green beans,** see Group 06: Legume vegetables

VD 4493   **Green gram (dry),** see Mung bean (dry)

VD 4495   **Groundnut,** see Peanut, Group 023: Oilseed

VD 4497   **Haricot bean,** see Common bean, Group 06: Legume vegetables

VD 4499   **Horse bean (dry),** see Broad bean (dry)

VD 562   **Horse gram**
           Dolichos uniflorus Lam.;
           syn: D. *biflorus* auct. non L.

VD 531   **Hyacinth bean (dry)**
           Lablab niger Medik
           syn: *Dolichos* lablab 1 .; Lablab *vulgaris* Savi.

VD 4501   **Jack bean,** see Group 06: Legume vegetables

VD 563   **Kersting's groundnut**
           Macrostyloma geocarpum (Harms) Marcechal & Baudet;
           syn: Kerstingiella *geocarpa* Harms.   Voandzeia *poissoinii* Chev.

VD 4503   **Kidney bean (dry),** see Common bean (dry)

VD 4505   **Lablab (dry),** see Hyacinth bean (dry)

VD 533   **Lentil (dry)**
           Lens esculenta Moench;
           syn: L. *Culinaris* Medik; *Ervum lens* L.

VD 534   **Lima bean (dry)**
           Phaseolus lunatus L.;
           syn: Ph. *limensis* Macf.; Ph *inamoenus* L.

VD 545   **Lupin (dry)**
           Lupinus spp., sweet spp. varieties and cultivars with a low alkaloid content

VD 535   **Mat bean (dry)**
           Phaseolus aconitifolius Jacq.

VD 4507   **Moth bean (dry),** see Mat bean (dry)

VD 536   **Mung bean (dry)**
           Phaseolus aureus Roxb;
           syn: *Vigna radiata* (L.) Wilczek, var. *radiata*;V. *aureus* (Roxb.) Hepper

VD 4509   **Navy bean (dry)**, see Common bean (dry)

VD 4511   **Pea (dry)**, see Field pea (dry)

VD 537   **Pigeon pea (dry)**
               Cajanus cajan (L.) Millsp.;
               syn: C. *indicus* Spreng.

VD 4513   **Red gram (dry)**, see Pigeon pea (dry)

VD 539   **Rice bean (dry)**
               Vigna umbellata (Thunb.) Ohwi & Ohashi;
               syn: V. *calcarata* (Roxb.) Kurz; *Phaseolus calcaratus* Roxb.

VD 4515   **Runner bean, see Common bean**, Group 06: Legume vegetables

VD 4517   **Scarlet runner bean,** see Group 06: Legume vegetables

VD 4519   **Sieva bean (dry)**, see Lima bean (dry)

VD 541   **Soya bean (dry)**
               Glycine max (L.) Merr.
               syn: G. soja Sieb. & Zucc.: G. *hispida* (Moench) *Maxim.*;
               Soja *max* (L.) Piper

VD 4521   **Soybean (dry)**, see Soya bean (dry)

VD 564   **Tepary bean (dry)**
               Phaseolus acutifolius Gray, var. *latifolius* Freem.

VD 4523   **Urd bean (dry)**, see Black gram (dry)

VD 4525   **Wrinkled pea (dry)**, see Field pea (dry)

Root and tuber vegetables

<u>Class A</u>
Type 2   Vegetables Group 016 Group Letter Code VR

Group 016. Root and tuber vegetables are the starchy enlarged solid roots, tubers, corms or rhizomes, mostly subterranean, of various species of plants, mostly annuals.

The underground location protects the edible portion from pesticides applied to the aerial parts of the crop during the growing season; however the commodities in this group are exposed to pesticide residues from soil treatments.

The entire vegetable may be consumed in the form of fresh or processed foods.

<u>Portion of the commodity to which the MRL applies (and which is analyzed)</u>: *Whole commodity after removing tops. Remove adhering soil (e.g. by rinsing in running water or by gentle brushing of the dry commodity).*

Group 016   Root and tuber vegetables

<u>**Code No.**</u>   <u>**Commodity**</u>

VR   75   **Root and tuber vegetables**

VR 4527   **Achira,** see Canna, edible

VR 570   **Alocasia**
                 Alocasia macrorrhiza (L.) Schott;
                 A. *indica* (Roxb.) Schott

VR 571   **Arracacha**
                 Arracacia xanthorrhiza Bancr.;
                 syn: A. *esculenta* DC.

VR 572   **Arrowhead**
                 Sagittaria sagittifolia L.; S. *sagittifolia* L.,
                 var. *sinensis* Sims;
                 S. *japonica* Hort.; S. *latifolia* Willd.;
                 S. *trifolia* L.; S. *trifolia* L., var. *edulis* Ohwi

VR 573   **Arrowroot**
                 Maranta arundinacea L.; several cultivars

VR 574   **Beetroot**
            Beta vulgaris L., var. *conditiva*

VR 4529   **Black salsify,** see Scorzonera

VR 575   **Burdock, greater or edible**
            Arctium lappa L.;
            syn: *Lappa officinalis* All.: L. major Gaertn.

VR 576   **Canna, edible**
            Canna edulis Ker.

VR 577   **Carrot**
            Daucus carota L.

VR 463   **Cassava**
            Manihot esculenta Crantz;
            syn: M. *aipi* Pohl; M. *ultissima* Pohl; M. *dulcis* Pax:
            M. *palmata* Muell.-Arg.

VR 4531   **Cassava, Bitter, see Cassava**
            Manihot esculenta Crantz, bitter cultivars

VR 4533   **Cassava, Sweet,** see Cassava
            Manihot esculenta Crantz, sweet cultivars

VR 578   **Celeriac**
            Apium graveolens L., var. *rapaceum* (Mill.) Gaudin

VR 423   **Chayote root**
            Sechium edule (Jacq.) Swartz

VR 579   **Chervil, Turnip-rooted**
            Chaerophyllum bulbosum L.

VR 469   **Chicory, roots**
            Cichorum intybus L. var *foliosum hegi* and var *sativum*
            Lam. & DC.

VR 4535   **Chinese radish,** see Radish, Japanese

VR 4537   **Christophine,** see Chayote root

VR 580   **Chufa,** see Tiger nut

VR 4539   **Cocoyam,** see Tannia and Taro

VR 4541   **Dasheen,** see Taro

VR 4543   **Daikon,** see Radish, Japanese

VR 4545   **Eddoe,** see Taro
            Colocasia esculenta L., var. *antiquorum* (Schott),
            Hubbard & Rehder;
            syn: *C. esculenta*, var. *globifera* Engl. & Krause

VR 581   **Galangal, Greater**
            Languas galanga (L.) Stuntz;
            syn: *Alpinia galanga* Sw.

VR 582   **Galangal, Lesser**
            Languas officinarum (Hance) Farwell:
            syn: *Alpinia officinarum* Hance

VR 4547   **Globe artichoke,** see Group 09: Artichoke Globe, Stalk and stem vegetables

VR 530   **Goa bean root**
            Psophocarpus tetragonolobus (L.) DC.

VR 4549   **Gruya,** see Canna, edible

VR 583   **Horseradish**
            Armoracia rusticana (Gaertn.) M. et Sch.;
            syn: *Cochlearia armoracia* L.:
            Armoracia lapathifolia Gilib.

VR 584   **Japanese artichoke**
            Stachys sieboldii Miq.

VR 585   **Jerusalem artichoke**
            Helianthus tuberosus L.

VR 4551   **Jicama,** see Yam bean

VR 4553   **Leren,** see Topee Tambu

VR 4555   **Manioc,** see Cassava, bitter

VR 586   **Oca**
            Oxalis tuberosa Mol.

VR 4557   **Oyster plant,** see Salsify

VR 587   **Parsley, Turnip-rooted**
            Petroselinum crispum (Mill.) Nyman ex A.W. Hill
            var. *tuberosum*

VR 588   **Parsnip**
            Pastinaca sativa L.

VR 589   **Potato**
            Solanum tuberosum L.

VR 4559   **Potato yam,** see Yam bean

VR 4561   **Queensland arrowroot,** see Canna, edible

VR 494   **Radish**
            Raphanus sativus L., Subvar. *radicola* Pers.

VR 590   **Radish, Black**
            Raphanus sativus L., subvar. *niger* Pers.

VR 591   **Radish, Japanese**
            Raphanus sativus L., var. *longipinnatus* Bailey

VR 592   **Rampion roots**
            Campanula rapunculus L.

VR 4563   **Rutabaga,** see Swede

VR 4564   **Red beet,** see Beetroot

VR 498   **Salsify**
            Tragopogon porrifolius L.

VR 4565   **Salsify, Black,** see Scorzonera

VR 593   **Salsify, Spanish**
            Scolymus hispanicus L.

VR 594   **Scorzonera**
           Scorzonera hispanica L.

VR 595   **Skirrit or Skirret**
           Sium sisarum L.

VR 596   **Sugar beet**
           Beta vulgaris L., var. *sacharifera*;
           syn: B. *vulgaris* L. var. *altissima*

VR 497   **Swede**
           Brassica napus L., var. *napobrassica* (L.) Reichenbach

VR 508   **Sweet potato**
           Ipomoea batatas (L.) Poir

VR 4567   **Tanier,** see Tannia

VR 504   **Tannia**
           Xanthosoma sagittifolium (L.) Schott

VR 4569   **Tapioca,** see Cassava

VR 505   **Taro**
           Colocasia esculenta (L.) Schott, var. *esculenta*

VR 580   **Tiger nut**
           Cyperus esculentus L.

VR 589   **Topee tambu**
           Calathea allouia (Aubl.) Lindl.

VR 4571   **Turnip,** see Swede

VR 506   **Turnip,** Garden
           Brassica rapa L., var. *rapa*;
           syn: B. *campestris* L., var. *rapifera*

VR 4573   **Turnip, Swedish,** see Swede

VR 599   **Ullucu**
           Ullucus tuberosus Caldas

VR 600   **Yams**
            Dioscorea L.: several species

VR 4575   **Yam, Cush-cush,** see Yams
            Dioscorea trifida L.

VR 4577   **Yam, Eight-months,** see Yam, White Guinea

VR 4579   **Yam, Greater,** see Yams
            Dioscorea alata L.

VR 4583   **Yam, Twelve-months,** see Yam, Yellow Guinea

VR 4585   **Yam, White,** see Yam, White Guinea

VR 4587   **Yam, White Guinea,** see Yams
            Dioscorea rotundata Poir.

VR 4589   **Yam, Yellow,** see Yam, Yellow Guinea

VR 4591   **Yam, Yellow Guinea,** see Yams
            Dioscorea cayenensis Lam.

VR 601   **Yam bean**
            Pachyrhizus erosus (L.) Urban;
            syn: P. *angulatus* Rich. ex DC.; P. *bulbosus* (L.) Kurz;
            Dolichos erosus L.

VR 4593   **Yautia,** see Tannia

Stalk and stem vegetables

<u>Class A</u>
Type 2   Vegetables Group 017 Group Letter Code VS

Group 017. Stalk and stem vegetables are the edible stalks, leaf stems or immature shoots, from a variety of annual or perennial plants. Although not actually belonging to this group, globe artichoke (the immature flowerhead) of the family *Compositae* is included in this group.

Depending upon the part of the crop used for consumption and the growing practices, stalk and stem vegetables are exposed, in varying degrees, to pesticides applied during the growing season.

Stalk and stem vegetables may be consumed in whole or in part and in the form of fresh, dried or processed foods.

<u>Portion of the commodity to which the MRL applies (and which is analyzed)</u>: *Whole commodity as marketed after removal of obviously decomposed or withered leaves. Rhubarb, leafstems only: globe artichoke, flowerhead only, celery and asparagus, remove adhering soil.*

Group 017   Stalk and stem vegetables

<u>Code No.</u>	<u>Commodity</u>
VS 78	**Stalk and stem vegetables**
VS 620	**Artichoke, Globe** *Cynara scolymus* L.
VS 621	**Asparagus** *Asparagus officinalis* L.
VS 622	**Bamboo shoots** *Bambusa vulgaris* Schrd. ex Wendland.; *Dendrocalamus strictus* (Roxb.) Nees *Gigantochloa verticilliata* (Willd.) Munro
VS 623	**Cardoon** *Cynara cardunculus* L.
VS 624	**Celery** *Apium graveolens* L., var. *dulce*
VS 4595	**Celery leaves,** see Group 027: Herbs

VS 625 **Celtuce**

Lactuca sativa L., var. *angustina* Irish;
syn: L. *sativa* L., var. *asparagina* Bailey

VS 626 **Palm hearts**

among others *Raphia* spp.; *Cocus nucifera* L.; *Borassus*
aethiopicum Mart.; Salacca *edilis* Reinw.

VS 627 **Rhubarb**

Rheum rhaponticum L.

VS 469 **Witloof chicory (sprouts)**

Cichorium intybus L., var. *foliosum* Hegi: green, red and white cultivars

TYPE 3   GRASSES

Grasses are herbaceous annual and perennial monocotyledonous plants of different kinds, cultivated extensively for their ears (heads) of starchy seeds used directly for the production of food.   Grasses used for animal feed are classified under Class C; Primary Animal feed commodities, Group 051.

The plants are fully exposed to pesticides applied during the growing season.

Cereal grains

<u>Class A</u>

Type 3   Grasses Group O20 Group Letter Code GC

Group 020. Cereal grains are derived from the ears (heads) of starchy seeds produced by a variety of plants, primarily of the grass family (*Gramineae*)

Buckwheat, a dicotyledonous crop belonging to the botanical family *Polygonaceae* and two *Chenopodium* species, belonging to the botanical family *Chenopodiaceae* are included in this group, because of similarities in size and type of the seed, residue pattern and the use of the commodity.

The edible seeds are protected to varying degrees from pesticides applied during the growing season by husks. Husks are removed before processing and/or consumption.

Cereal grains are often exposed to post-harvest treatments with pesticides.

<u>Portion of the commodity to which the MRL applies (and which is analyzed)</u>: *Whole commodity. Fresh corn and sweet corn: kernels plus cob without husk. (For the latter group see 012 Fruiting vegetables, other than Cucurbits).*

For Fodders and straw of cereals, see Class C, Type 11 Group 051.

Group 020   Cereal grains

<u>Code No.</u>   <u>Commodity</u>

GC   80   **Cereal grains**
             Seeds of *gramineous* plants as listed below, and Buckwheat and *Chenopodium* spp. as listed

GC   81   **Cereal grains, except Buckwheat, Cañihua and Quinoa**

GC 4597   **Acha,** see Hungry Rice

GC 4599   **Adlay,** see Job's Tears

GC 4601   **African millet,** see Millet, Finger

GC 640   **Barley**
    Hordeum vulgare L.;
    syn: H *sativum* Pers.

GC 4603   **Brown-corn millet,** see Millet, Common

GC 641   **Buckwheat**
    Fagopyrum esculentum Moench;
    syn: F. *sagittatum* Gilib.

GC 4607   **Bulrush millet,** see Millet, Bulrush

GC 642   **Cañihua**
    Chenopodium pallidicaule Aellen

GC 4609   **Cat-tail millet,** see Millet, Bulrush

GC 4611   **Chicken corn,** see Sorghum
    Sorghum drummondii (Steud.) Millsp. & Chase

GC 4613   **Corn, see Maize**

GC 4615   **Corn-on-the-cob** (Codex Stan. 133-1981), see Fruiting vegetables (other than Cucurbits), Group 04

GC 4617   **Corn, whole kernel** (Codex Stan. 132-1981, see Fruiting vegetables (other than Cucurbits), Group 04: Sweet corn (kernels)

GC 4619   **Dari seed,** see Sorghum

GC 4621   **Durra,** see Sorghum
    ssp. *Sorghum durra* (Forsk.) Stapf.

GC 4623   **Durum wheat,** see Wheat
    ssp. *Triticum durum* Desf.

GC 4625   **Emmer,** see Wheat
    ssp. *Triticum dicoccum* Schubl.

GC 4627   **Feterita,** see Sorghum
                   ssp. *Soryhum caudatum* Stapf.

GC 4629   **Finger millet,** see Millet, Finger

GC 4631   **Fonio,** see Hungry Rice

GC 4633   **Foxtail millet,** see Millet, Foxtail

GC 4635   **Fundi,** see Hungry Rice

GC 4637   **Guinea corn,** see Sorghum
                   spp. *Sorghum guineense* Stapf.

GC 4639   **Hog millet,** see Millet, Common

GC 643   **Hungry rice**
                   Digitaria exilis Stapf.; D. *iburua* Stapf.

GC 644   **Job's tears**
                   Coix *lacryma*-jobi L.

GC 4641   **Kaffir corn,** see Sorghum
                   ssp. *Sorghum caffrorum* Beauv.

GC 4643   **Kaoliang,** see Sorghum
                   ssp. *Sorghum nervosum* Bess. ex Schult

GC 645   **Maize**
                   Zea mays L., several cultivars, not including Popcorn and Sweetcorn

GC 646   **Millet**
                   Including Barnyard Millet, Bulrush Millet, Common Millet, Finger Millet,
                   Foxtail Millet, Little Millet;
                   see for scientific names, specific commodities listed as Millet, followed by a
                   specific denomination

GC 4645   **Millet, Barnyard,** see Millet
                   Echinochloa crus-galli (L.) Beauv.;
                   syn: *Panicum* crus-galli L.:
                   E. *frumentacea* (Roxb.) Link;
                   syn: *Panicum frumentaceum* Roxb.

GC 4647 **Millet, Bulrush,** see Millet
Pennisetum typhoides (Burm.f.) Stapf. & Hubbard;
syn: P. *glaucum* (L.) R. Br. P. *americanum* (L.) K.Schum.:
P. *spicatum* (L.) Koern.

GC 4649 **Millet, Common,** see Millet
Panicum miliaceum L.

GC 4651 **Millet, Finger,** see Millet
Eleusine coracana (L.) Gaertn.

GC 4653 **Millet, Foxtail,** see Millet
Setaria italica (L.) Beauv
Syn: *Panicum italicum* L. *Chaetochloa italica* (L.) Scribn.

GC 4655 **Millet, Little,** see Millet
Panicum sumatrense Roth ex Roem & Schult.

GC 4657 **Milo,** see Sorghum
ssp. *Sorghum subglabrescens* Schweinf. & Aschers

GC 647 **Oats**
Avena fatua L.; A. *abyssinica* Hochst.

GC 4659 **Oat, Red,** see Oats
Avena byzantina Koch

GC 4661 **Pearl millet,** see Millet, Bulrush

GC 656 **Popcorn**
Zea mays L., var. *everta* Sturt.
syn: *Zea mays* L., var. *praecox*

GC 4665 **Proso millet,** see Millet, Common

GC 648 **Quinoa**
Chenopodium quinoa Willd.

GC 649 **Rice**
Oryza sativa L.; several ssp. and cultivars

GC 4667 **Russian millet,** see Millet, Common

GC 650   **Rye**
           Secale cereale L.

GC 4669   **Shallu,** see Sorghum
           ssp. *Sorghum roxburghii* Stapf.

GC 4671   **Sorgo,** see Sorghum

GC 651   **Sorghum**
           Sorghum bicolor (L.) Moench; several Sorghum ssp. and cultivars

GC 4673   **Spelt,** see Wheat
           Triticum spelta L.

GC 4675   **Spiked millet,** see Millet, Bulrush
           Sweet corn, see Group 04 Fruiting vegetables (other than Cucurbits)

GC 652   **Teff or Tef**
           Eragrostis tef (Zucc.) Trotter;
           syn: E. *abyssinica* (Jacq.) Link

GC 657   **Teosinte**
           Zea mays ssp. *mexicana* (Schrader) *Iltis*:
           syn: Zea *mexicana* (Schrader) Kunze: *Euchlaena mexicana*
           Schrader.

GC 653   **Triticale**
           Hybrid of Wheat and Rye

GC 654   **Wheat**
           Cultivars of *Triticum aestivum* L.:
           syn: T. *sativum* Lam.: T. *vulgare* Vill.; *Triticum* spp., as listed

GC 655   **Wild rice**
           Zizania aquatica L.

Grasses for sugar or syrup production

<u>Class A</u>

Type 3   **Grasses Group 021 Group Letter Code GS**

Group 021. Grasses For sugar or syrup production, includes species of grasses with a high sugar content especially in the stem. The stems are mainly used for sugar or syrup production, and to a small extent as vegetables or sweets. The leaves, ears and several wastes of the sugar or syrup manufacturing process are used, among others, as animal feed see Group 052: Miscellaneous fodder and forage crops).

Group 021   Grasses for sugar or syrup production

<u>Code No.</u>   <u>Commodity</u>

GS 658   **Sorgo or Sorghum, Sweet**
             Varieties and cultivars of Sorghum of which the stems contain considerable amounts of sweet juice among others cultivars of *Sorghum bicolor* (L.) Moench. S. *dochnu* (Forsk.) Snowdon

GS 659   **Sugar cane**
             Saccharum officinarum L.

TYPE 4   NUTS AND SEEDS

Nuts and seeds are derived from a large variety of trees, shrubs and herbaceous plants, mostly cultivated.

The mature seeds or nuts are used as human food, for the production of beverages or edible vegetable oils and for the production of seed meals and cakes for animal feed.

Tree nuts

Class A

Type 4   Nuts and seeds Group 022 Group Letter Code TN

Group 022. Tree nuts are the seeds of a variety of trees and shrubs which are characterized by a hard inedible shell enclosing an oily seed.

The seed is protected from pesticides applied during the growing season by the shell and other parts of the fruit.

The edible portion of the nut is consumed in succulent, dried or processed forms.

Portion of the commodity to which the MRL applies (and which is analyzed): *Whole commodity after removal of shell. Chestnuts: whole in skin.*

Group 022   Tree nuts

Code No.	Commodity

TN 85   Tree nuts

TN 660   Almonds
Prunus dulcis (Mill.) D.A. Webb, syn: *Amygdalus communis* L., *Prunus amygdalus* Batsch.

TN 661   Beech nuts
Fagus sylvatica L.; F. *grandifolia* Ehrh.

TN 662   Brazil nut
Bertholletia excelsa Humb. & Bonpl.

TN 4681   Bush nut, see Macadamia nut

TN 663   Butter nut
Juglans cinerea L.

TN 295   **Cashew nut**
            Anacardium occidentale L.

TN 664   **Chestnuts**
            Castanea sativa Mill., syn: C. *vesca* Gaertn.
            Castanea molissima Blume; *Castanea pumila* (L.) Mill.

TN 4683   **Chinquapin,** see Chestnuts
            Castanea pumila (L.) Mill.

TN 665   **Coconut**
            Cocos nucifera L.

TN 4685   **Filberts,** see Hazelnuts
            among others *Corylus maxima* Mill.

TN 666   **Hazelnuts**
            Corylus avellana L.; C. *maxima* Mill.

TN 667   **Hickory nuts**
            Carya ovata Koch.; C. glabra (Mill.) other sweet Carya species

TN 668   **Japanese horse-chestnut**
            Aesculus turbinata Blume;
            syn: Ae. *sinensis* Hort., not bunge

TN 4687   **Java almonds,** see Pili nuts
            Canarium commune L.; C. *indicum* L.; C. *amboinensis* Hochst.; C.
            moluccanum Blume

TN 669   **Macadamia nuts**
            Macadamia ternifolia F. Muell.: M. *tetraphylla* (L.) Johnson

TN 670   **Pachira nut**
            Pachira insignis Savigny

TN 671   **Paradise nut,** see Sapucaia nut
            Lecythis zabucajo Aubl.

TN 672   **Pecan**
            Carya illinoensis (Wangh) K. Koch

TN 4689   **Pignolia or Pignoli,** see Pine nuts

TN 673   **Pine nuts**
          Mainly *Pinus pinea* L.; also P. *lambertiana* Dougl.; P *cembra* L.; P *edulis*
          Engelm.; P *quadrifolia* Parl. ex Sudw. and other *Pinus* species

TN 4691   **Pinocchi,** see Pine nuts

TN 4693   **Piñon nut,** see Pine nuts

TN 674   **Pili nuts**
          Canarium ovatum Engl.; C. *luzonicum* A Gray; C. *pachyphyllum* Perkins; C.
          commune L.

TN 675   **Pistachio nut**
          Pistachia vera L.

TN 4695   **Queensland Nut,** see Macadamia nut

TN 676   **Sapucaia** nut
          Lecythis zabucajo Aubl.; L. *elliptica* Kunth; L. *ollaria* L.: L. *usitatis* Miers

TN 677   **Tropical almond**
          Terminalia catappa L.

TN 678   **Walnuts**
          Juglans *regia* L.; J. *nigra* L.

TN 4697   **Walnut, Black,** see Walnuts
          Juglans *nigra* L

TN 4699   **Walnut, English; Walnut, Persian,** see Walnuts
          Juglans *regia* L.

Oilseed

<u>Class A</u>
Type 4   Nuts and seeds   Group 023   Group Letter Code S0

Group 023.   Oilseed consists of seeds from a variety of plants used in the production of edible vegetable oils, seed meals and cakes for animal feed.   Some important vegetable oil seeds are by-products of fibre or fruit crops (e.g. cotton seed, olives).

Some of the oilseeds are, directly or after slight processing (e.g. roasting), used as food (e.g. peanuts) or for food flavouring (e.g. poppy seed, sesame seed).

Oilseeds are protected from pesticides applied during the growing season by the shell or husk.

<u>Portion of the commodity to which the MRL applies (and which is analyzed)</u>: *Unless specified, seed or kernels, after removal of shell or husk.*

Group 023   Oilseed

<u>Code No.</u>	<u>Commodity</u>
SO 88	**Oilseed**
SO 89	**Oilseed except peanut**
SO 90	**Mustard seeds** (Mustard seed; Mustard seed, Field: Mustard seed, Indian)
SO 690	**Ben Moringa seed** Moringa *oleifera* Lam., syn: M. *pterygosperma* Gaertn.; M. *perigrina* (Forsk.) *Fiori*
SO 4701	**Coconut,** see Group 022: Tree nuts
SO 4703	**Colza,** see Rape seed
SO 4705	**Colza, Indian,** see Mustard seed, Field *Brassica campestris* L., var. sarson Prain
SO 691	**Cotton seed** *Gossypium* spp.; several species and cultivars
SO 4707	**Desert date,** see Group 5: Assorted tropical and sub-tropical fruits - edible peel

SO 4709   **Drumstick tree seed,** see Ben Moringa seed

SO 4711   **Flax-seed,** see Linseed

SO 4713   **Groundnut,** see Peanut

SO 4715   **Horseradish tree seed,** see Ben Moringa seed

SO 692   **Kapok**
               Ceiba pentandra (L.) Gaertn.

SO 693   **Linseed**
               Linum usitatissimum L.

SO 4718   **Maize,** see Group 020: Cereal Grains

SO 485   **Mustard seed**
               Brassica nigra (L.) Koch
               Sinapis alba L., syn: *Brassica hirta* Moench.

SO 694   **Mustard seed,** Field
               Brassica campestris L., var. sarson Prain:
               B. *campestris* L., var. *toria* Duthie & Fuller

SO 478   **Mustard seed, Indian**
               Brassica juncea (L.) Czern. & Coss.

SO 695   **Niger seed**
               Guizotia abyssinica (L.) Cass.

SO 4719   **Olive,** see Group 5: Assorted tropical and sub-tropical fruits - edible peel

SO 696   **Pala nut**
               Elaeis guineensis Jacq.

SO 697   **Peanut**
               Arachis hypogaea L.

SO 703   **Peanut, whole**

SO 698   **Poppy seed**
               Papaver somniferum L.

SO 495   **Rape seed**
            Brassica napus L.

SO 4721   **Rape seed, Indian,** see Mustard seed, Field
            Brassica campestris L., var. *toria* Duthie & Fuller

SO 699   **Safflower seed**
            Carthamus tinctorius L.

SO 700   **Sesame seed**
            Sesamum indicum L.;
            syn: *S. orientale* L.

SO 701   **Shea nuts**
            Butyrospermum paradoxum (Gaertn.) Hepper, subsp. *parkii* (G. Don.) Hepper;
            syn: B. *parkii* (G. Don.) Kotsky

SO 4723   **Soya bean (dry),** see Group 07: Pulses

SO 4724   **Soybean (dry),** see Soya bean (dry)

SO 702   **Sunflower seed**
            Helianthus annuus L.

Seed for beverages and sweets

<u>Class A</u>
Type 4   Nuts and seeds Group 024 Group Letter Code SB

The seeds for beverages and sweets are derived from tropical and sub-tropical trees and shrubs. After processing the seeds are used in the production of beverages and sweets.

These seeds are protected from pesticides applied during the growing season by the shell or other parts of the fruit.

<u>Portion of the commodity to which MRL applies (and which is analyzed)</u>: *Unless specified, whole commodity (seed only, other parts of the fruit not included).*

Group 024   Seed for beverages and sweets

<u>Code No.</u>	<u>Commodity</u>
SB 91	**Seed for beverages**
SB 715	**Cacao beans** *Theobroma cacao* L.; several ssp.
SB 716	**Coffee beans** among others *Coffea arabica* L. C. *canephora* Pierre ex Froehner C. *liberica* Bull ex Hiern.; ssp. and cultivars
SB 717	**Cola nuts** *Cola nitida* (Vent.) Schott & Endl.; C. *acuminata* (P. Beauv.) Schott & Endl.; C. *anomala* K. Schum.; C. *verticillata* (Thonn.) Stapf ex A. Chev.
SB 4727	**Kola,** see Cola nuts

TYPE 5   HERBS AND SPICES

Herbs and spices are the flavoursome or aromatic leaves, stems, roots, flowers or fruits of a variety of plants used to impart special flavours to food and beverages.

Herbs

<u>Class A</u>
Type 5   **Herbs and spices Group 027 Group Letter Code HH**

Herbs consist of leaves, flowers, stems and roots from a variety of herbaceous plants, used in relatively small amounts as condiments to flavour foods or beverages. They are used either in fresh or naturally dried form.

Herbs are fully exposed to pesticides applied during the growing season. Post-harvest treatments are often carried out on dried herbs.

Herbs are consumed as components of other foods in succulent and dried forms or as extracts of the succulent products.

<u>Portion of the commodity to which the MRL applies (and which is analyzed)</u>: **Whole commodity as prepared for wholesale or retail distribution.**

Group 027   Herbs

<u>Code No.</u>	<u>Commodity</u>
HH 726	**Herbs**
HH 720	**Angelica, including Garden Angelica** *Angelica sylvestris* L.; A. *archangelica* L.
HH 721	**Balm leaves** *Melissa officinalis* L.
HH 722	**Basil** *Ocimum basilicum* L.
HH 723	**Bay leaves** *Laurus nobilis* L.
HH 724	**Borage** *Borago officinalis* L.

HH 725   **Burnet, Great**
            Sanguisorba officinalis L.;
            syn: *Poterium officinala* A. Gray

HH 4731   **Burnet, Salad,** see Burnet, Great
            Sanguisorba minor Scop.;
            syn: *Poterium sanguisorba* L.

HH 720   **Burning bush**
            Dictamnus albus L.:
            syn: D. *fraxinella* Pers.

HH 726   **Catmint**
            Nepeta cataria L.

HH 4733   **Catnip,** see Catmint

HH 624   **Celery leaves**
            Apium graveolens L.; var. *seccalinum* Alef

HH 4735   **Chervil,** see Group 013: Leafy vegetables

HH 727   **Chives**
            Allium schoenoprasum L.

HH 4737   **Chives, Chinese,** see Chives
            Allium tuberosum Rottl. ex Spreng.;
            syn: A *odoratum* L.

HH 4739   **Clary,** see Sage (and related Salvia species)
            Salvia sclarea L.

HH 4741   **Costmary,** see Tansy (and related species)
            Tanacetum balsamita L.;
            syn: *Chrysanthemum balsamita* L.

HH 4743   **Cretan Dittany,** see Burning bush

HH 729   **Curry leaves**
            Murraya koenigii (L.) Spreng.

HH 730   **Dill**
            Anethum graveolens L.

HH 4745   **Estragon,** see Tarragon

HH 731   **Fennel**
             Foeniculum vulgare Mill.;
             syn: *F. officinale* All.; *F capillaceum* Gilib.

HH 4747   **Fennel, Bulb,** see Group 01: Bulb vegetables, No. VA 0380

HH 732   **Horehound**
             Marrubium vulgare L.

HH 733   **Hyssop**
             Hyssopus officinalis L.

HH 734   **Lavender**
             Lavendula angustifolia Mill.;
             syn: *L. officinalis* Chaix; *L. spica* L.; *L. vera* DC.

HH 735   **Lovage**
             Levisticum officinale Koch.

HH 737   **Marigold flowers**
             Calendula officinalis L.

HH 736   **Marjoram**
             Origanum marjorana L.;
             syn: *Marjorana hortensis* Moench.;
             Origanum vulgare L.

HH 4749   **Marjoram, Sweet,** see Marjoram
             Marjorana hortensis Moench.;
             syn: *Origanum marjorana* L.

HH 4751   **Marjoram, Wild,** see Marjoram
             Origanum vulgare L.

HH 738   **Mints**
             Several Mentha species and hybrids; (see also individual Mints) including
             Mentha spicata L.; *M. spicata* L., var. *crispata* (Schrad.) Sch. et Thell.; *M. x
             piperata* L.; *M. x gentilis*

HH 4753   **Mugwort,** see Wormwoods
             Artemisia vulgaris L.

HH 4755   **Myrrh,** see Sweet Cicely

HH 739   **Nasturtium, Garden, leaves**
    Tropaeolum majus L.

HH 4757   **Oregano,** see Marjoram

HH 740   **Parsley**
    Petroselinum crispum (Mill.) Nyman ex A.W. Hill;
    syn: P. *sativum* Hoffm.; P. *hortense* auct.

HH 4759   **Pennyroyal,** see Mints
    Mentha pulegium L.;
    syn: *Pulegium vulgare* Mill.

HH 4761   **Peppermint,** see Mints
    Mentha x piperita L. (= hybrid of M. *aquatica* L. x M. spicata L.)

HH 741   **Rosemary**
    Rosmarinus officinalis L.

HH 742   **Rue**
    Ruta graveolens L.

HH 743   **Sage and related Salvia species**
    Salvia officinalis L.; *S. sclarea* L.

HH 744   **Sassafras leaves**
    Sassafras albidum Nees

HH 745   **Savory, Summer; Winter**
    Satureja hortensis L.; *S montana* L.

HH 746   **Sorrel, Common, and related Rumex species**
    among others *Rumex acetosa* L.; *R. scutatus* L.: *R. patientia* L.; *R. rugosus*
    Campd.

HH 4763   **Southernwood,** see Wormwoods
    Artemisia abrotanum L.

HH 4765   **Spearmint,** see Mints
    Mentha spicata L.;
    syn: M. *viridis* L.

HH 747 **Sweet Cicely**
Myrrhis odorata (L.) Scop.

HH 748 **Tansy and related species**
Tanacetum vulgare L.;
T *balsamita* L.; syn: *Chrysanthemum balsamita* L.

HH 749 **Tarragon**
Artemisia dracunculus L.
including *Thymus vulgaris* L.; Th. *serpyllum* L. and *Thymus* hybrids.

HH 4767 **Watercress,** see Group 05: Leafy vegetables

HH 751 **Winter cress,** Common; American
Barbarea vulgaris R. Br.; B. *verna* (Mill.) Aschrs.

HH 752 **Wintergreen leaves**
Gaultheria procumbens L.
(not including herbs of the Wintergreen family *(Pyrolaceae)*)

HH 753 **Woodruff**
Asperula odorata L.

HH 754 **Wormwoods**
Artemisia absinthium L.; A. *abrotanum* L.; A *vulgaris* L.

Spices

<u>Class A</u>
Type 5   Herbs and spices Group 028 Group Letter Code HS

Group 028. Spices consist of the aromatic seeds, roots, berries or other fruits from a variety of plants, which are used in relatively small quantities to flavour foods.

Spices are exposed in varying degrees to pesticides applied during the growing season. Also post-harvest treatments may be applied to spices in the dried form.

They are consumed primarily in the dried form as condiment.

<u>Portion of the commodity to which MRL applies (and which is analyzed)</u>: *Unless specified, whole commodity as marketed, mainly in the dried form.*

Group 028   Spices

<u>Code No.</u>	<u>Commodity</u>
HS 93	**Spices**
HS 4769	**Allspice fruit, see Pimento**
HS 720	**Angelica seed** *Angelica archangelica* L.; *A. sylvestris* L.
HS 4771	**Angelica, root, stem and leaves,** see Group 027: Herbs, Angelica, including Garden Angelica
HS 4773	**Aniseed,** see Anise seed
HS 771	**Anise seed** *Pimpinella anisum* L.
HS 772	**Calamus, root** *Acorus calamus* L.
HS 773	**Caper buds** *Capparis spinosa* L.
HS 774	**Caraway seed** *Carum carvi* L.

HS 775   **Cardamom seed**
            Elettaria cardamomum Maton

HS 4775   **Cassia bark,** see Cinnamon bark (including Cinnamon, Chinese bark)

HS 776   **Cassia buds**
            Cinnamomum cassia (Nees) Nees ex Blume

HS 624   **Celery seed**
            Apium graveolens L.

HS 777   **Cinnamon bark** (including Cinnamon, Chinese bark)
            Cinnamomum zeylanicum Breyn; *C. cassia* (Nees) Nees ex Blume

HS 778   **Cloves, buds**
            Syzygium aromaticum (L.) Merr. & Perr.
            syn: *Eugenia caryophyllus* (Sprengel) Bullock & Harrison; *E. aromatica* Kuntze;
            E. caryophyllata Thunb.; *Caryophyllus aromaticus* L.

HS 779   **Coriander, seed**
            Coriandrum sativum L.

HS 780   **Cumin seed**
            Cuminum cyminum L.

HS 730   **Dill seed**
            Anethum graveolens L.

HS 781   **Elecampane, root**
            Inula helenium L.

HS 731   **Fennel, seed**
            Foeniculum vulgara Mill.;
            syn: F. *officinale* All; F *capilaceum* Gilib.

HS 782   **Fenugreek, seed**
            Trigonella foenum-graecum L.

HS 783   **Galangal, rhizomes**
            Languas galanga (L.) Stunz; syn: *Alpinia galanga* Sw.; *Languas officinarum*
            (Hance) Farwell; syn: *Alpinia officinarum* Hance

HS 784   **Ginger, root**
            Zingiber officicinale Rosc.

HS 785   **Grains of paradise**
Aframonum melegueta (Rosc.) K Schum.;
syn: *Amomum melegueta* Rosc.

HS 4779   **Horseradish,** see VR 0583, Group 016: Root and Tuber vegetables

HS 786   **Juniper, berry**
Juniperis communis L.

HS 4781   **Licorice,** see Liquorice

HS 787   **Liquorice, roots**
Glycyrrhiza glabra L.

HS 735   **Lovage, seed**
Levisticum officinale Koch.

HS 788   **Mace**
Dried aril of *Myristica fragrans* Houtt.

HS 739   **Nasturtium pods**
Tropaeolum majus L.

HS 789   **Nutmeg**
Seed of *Myristica fragrans* Houtt.

HS 790   **Pepper, Black; White** (see Note)
Piper nigrum L.

HS 791   **Pepper, Long**
Piper longum L.; *P. retrofractum* Vahl.;
syn: *P. officinarum* DC.

HS 792   **Pimento, fruit**
Pimenta dioica (L.) Merrill;
syn: *P. officinalis* Lindl.

HS 4783   **Poppy seed,** see Group 023: Oilseed

HS 4785   **Sesame seed,** see Group 023: Oilseed

HS 4787   **Tamarind,** see Group 6: Assorted tropical and sub-tropical fruits - inedible peel

HS 370   **Tonka bean,** see also Group 6: Assorted tropical and sub-tropical fruits - inedible
            peel
            Dipteryx odorata (Aubl.) Willd.

HS 794   **Turmeric, root**
            Curcuma domestica Val.;
            syn: C. *longa* Keunig non L.

HS 795   **Vanilla, beans**
            Vanilla mexicana Mill.:
            syn: V. *fragrans* (Salisb.) Ames, V. *planifolia* Andrews

<u>Note</u>: Although white pepper is in principle a processed food belonging to Type 13: Derived products of plant origin, it is listed for convenience in Group 028 Spices. White pepper is prepared from Black pepper, *Piper nigrum* L.; The seeds are wetted in water and dried after removal of the mesocarp. The resulting white pepper may or may not be ground into powder.

CLASS B   PRIMARY FOOD COMMODITIES OF ANIMAL ORIGIN

For the purpose of the Codex Alimentarius the term "primary food commodity" means the product in or nearly in its natural state, intended for the processing into food for sale to the consumer or intended for sale to the consumer as a food without further processing. It includes irradiated primary food commodities and products after removal of certain parts of the animal tissue, e.g. bones.

Food commodities of animal origin are parts of domesticated or wild animals, including their eggs and mammary secretions.

TYPE 6   MAMMALIAN PRODUCTS

Mammalian products are derived from the edible parts of various mammals, primarily herbivorous, slaughtered for food. These mammals are usually domesticated, or to a lesser extent obtained as game animals. This type does not include edible products from marine mammals, for which see Group 044.

Meat (from mammals other than marine mammals)

<u>Class B</u>
Type 6   Mammalian products Group 030 Group Letter Code MM

Group 030.   Meats are the muscular tissues, including adhering fatty tissues such as intramuscular and subcutaneous fat from animal carcasses or cuts of these as prepared for wholesale or retail distribution in a "fresh" state. The cuts offered to the consumer may include bones, connective tissues and tendons as well as nerves and lymph nodes.

The commodity description of "fresh" meat includes meat which has been quick-frozen or quick-frozen and thawed.

The Group does not include edible offal as defined in Group 032.

Exposure to pesticides is through animal metabolism following oral intake with feed or through dermal intake as a consequence of external use of pesticides against ectoparasites.

The entire commodity except bones may be consumed.

<u>Portion of commodity to which the MRL applies (and which is analyzed)</u>: *Whole commodity (without bones). For fat-soluble pesticides a portion of adhering fat is analyzed and MRLs apply to the fat. For those commodities where the adhering fat is insufficient to provide a suitable sample, the whole commodity (without bone) is analyzed and the MRL applies to the whole commodity (e.g. rabbit meat) (Ref. ALINORM 87/24, Appendix IV, paragraph 6).*

Group 030   **Meat** (from mammals other than marine mammals)

<u>Code No.</u>   <u>Commodity</u>

MM   95   **Meat** (from mammals other than marine mammals)

MM   96   **Meat of cattle, goats, horses, pigs and sheep**

MM   97   **Meat of cattle, pigs and sheep**

MM   810   **Buffalo meat**
                 Bubalus bubalis L.
                 Syncerus caffer Sparrman
                 Bison bison L.

MM   4789   **Buffalo, African, meat,** see Buffalo meat
                 Syncerus caffer Sparrman

MM   4791   **Buffalo, American, meat,** see Buffalo meat
                 Bison bison L.

MM   4793   **Buffalo, Cape, meat,** see Buffalo, African, meat

MM   4795   **Buffalo, Water, meat,** see Buffalo meat
                 Bubalus bubalis L.;
                 syn: *Bubalis buffalus* Blum,
                 Bos bubalis Brise *Bubalis bos* Wall.

MM   4797   **Calf meat,** see Cattle meat

MM   811   **Camel meat**
                 Camelus bactrianus L.; *C. dromedarius* L.;
                 Lama *glama* L.; Lama *pacos* L.

MM   4799   **Camel, Bactrian, meat,** see Camel meat
                 Camelus bactrianus L.

MM   812   **Cattle meat**
                 Breeds and interbreeds of among others *Bos taurus* L.: B *indicus* L.; B.
                 grunniens L.; breeds of B. *javanicus* d'Alton

MM   813   **Deer meat**
                 among others *Cervus elaphus germanicus* Desmarest: *Dama dama dama* L.;
                 syn: *Cervus dama* Corbet & Hill: other *Cervus* spp. and spp.

MM 4803   **Deer, Fallow, meat,** see Deer meat
Dama dama dama L.;
syn: *Cervus dama* Corbet & Hill

MM 4805   **Deer, Red, meat,** see Deer meat
Cerphus elaphus L.: other *Cerphus* spp., several ssp.

MM 4807   **Dromedary meat,** see Camel meat
Camelus dromedarius L.

MM 824   **Elk meat**
Alces alces (L.)

MM 814   **Goat meat**
Breeds of *Capra hircus* L. other *Capra* spp., several breeds.

MM 815   **Hare meat**
Lepus europaeus Pallas, several ssp. and var.;
L. *timidus* L., several var.; other *Lepus* spp.

MM 816   **Horse meat**
Several breeds of *Equus caballus* L.

MM 817   **Kangaroo meat**
Genera of the family *Macropodinae*

MM 4809   **Lamb meat,** see Sheep meat

MM 4811   **Llama or Lama meat,** see Camel meat
Breeds of *Lama glama* L.; *Lama pacos* L.

MM 4813   **Moufflon meat,** see sheep meat
Ovis musimon (Pallas);
syn: *Aegoceros musimon* Pallas

MM 4815   **Moose, European, meat,** see Elk meat

MM 818   **Pig meat**
among others *Sus domesticus Erxleben* and breeds; *Sus* spp. and ssp.

MM 819   **Rabbit meat**
Oryctolagus cuniculus L.
O. *cuniculus fodiens* (Gray); *Lepus cuniculus* L.; *Sylvilagus* spp.

MM 820 **Reindeer meat**
Rangifer tarandus (L.) (dom.)

MM 821 **Roe meat**
Capreolus capreolus capreolus (L.)

MM 822 **Sheep meat**
Several breeds of *Ovis aries* L.; other *Ovis* spp.

MM 4817 **Veal** (= calf meat), see Cattle meat

MM 4819 **Water Buffalo,** meat, see Buffalo meat
Bubalus bubalis L.

MM 823 **Wild boar meat**
Sus scrofa scrofa L.

MM 4821 **Yak meat,** see Cattle meat
Bos grunniens L.

MM 4823 **Zebu meat,** see Cattle meat
Bos indicus L.

Mammalian fats (except fat from marine mammals)

Class B
Type 6   Mammalian products Group 031 Group Letter Code MF

Group 031. Mammalian fats, excluding milk fats are derived from the fatty tissues of animals (not processed). For processed animal fats see Group 085.

Exposure to pesticides is through animal metabolism following oral intake with feed or through dermal intake as a consequence of external use of the pesticides against ectoparasites.

The entire commodity may be consumed.

Portion of the commodity to which the MRL applies (and which is analyzed): *Whole commodity.*

The scientific species names of the relevant animals are not repeated for this group of commodities. For these names see Group 030 Meat (from mammals other than marine mammals).

Group 031   Mammalian fats

Code No.	Commodity
MF 100	**Mammalian fats (except milk fats)**
MF 810	**Buffalo fat**
MF 811	**Camel fat**
MF 812	**Cattle fat**
MF 814	**Goat fat**
MF 815	**Hare fat**
MF 816	**Horse fat**
MF 818	**Pig fat**
MF 819	**Rabbit fat**
MF 822	**Sheep fat**

Edible offal (mammalian)

Class B

Type 6   Mammalian products Group 032   Group Letter Code MO

Group 032. Edible offal are edible tissue and organs other than muscles (= meat) and animal fat from slaughtered animals as prepared for wholesale or retail distribution.

Examples: liver, kidney, tongue, heart, stomach, sweetbread (thymus gland), brain, etc.

The group name and definitions are in conformity with those recorded in the Codex Standards 89-1981 and 98-1981, Codex Standard for Luncheon Meat and Codex Standard for Cooked Cured Chopped Meat respectively: "Edible offal" means such offal as have been passed as fit for human consumption, but not including lungs, ears, scalp, snout (including lips and muzzle), mucous membranes, sinews, genital system, udders, intestines and urinary bladder". In the former Classification of Food and Food Groups in the Guide to Codex Maximum Limits for Pesticide Residues 1978; CAC/PR 1-1978 the name Meat by-products was used for this group.

Exposure to pesticides is through animal metabolism following oral intake with feed or through dermal intake as a consequence of external use of pesticides on livestock animals against ectoparasites.

The entire commodity may be consumed.

Portion of the commodity to which the MRL applies (and which is analyzed): *Whole commodity.*

The scientific species names of the relevant animals are not repeated for this group of commodities. For these names see Group 030 Meat (from mammals other than marine mammals).

Group 032   Edible offal (Mammalian)

Code No.	Commodity
MO 105	**Edible offal (Mammalian)**
MO 96	**Edible offal of cattle, goats, horses, pigs and sheep**
MO 97	**Edible offal of cattle, pigs and sheep**
MO 98	**Kidney of cattle, goats, pigs and sheep**
MO 99	**Liver of cattle, goats, pigs and sheep**

MO 810	**Buffalo, Edible offal of**
MO 811	**Camel, Edible offal of**
MO 812	**Cattle, Edible offal of**
MO 1280	**Cattle, Kidney**
MO 1281	**Cattle, liver**
MO 14	**Goat, Edible offal of**
MO 816	**Horse, Edible offal of**
MO 1292	**Horse, kidney**
MO 1293	**Horse, liver**
MO 818	**Pig, Edible offal of**
MO 1284	**Pig, kidney**
MO 1285	**Pig, liver**
MO 822	**Sheep, Edible offal of**
MO 1288	**Sheep, kidney**
MO 1289	**Sheep, liver**

Milks

<u>Class B</u>
Type 6   Mammalian products Group 033 Group Letter Code ML

Group 033. Milks are the mammary secretions of various species of lactating herbivorous ruminant animals, usually domesticated.

In conformity with the Codex Alimentarius Code of Principles concerning Milk and Milk Products the term "Milk" shall mean exclusively the normal mammary excretion obtained from one or more milkings without either addition thereto or extraction therefrom.

Notwithstanding the provisions in the preceding paragraph, "the term "Milk" may be used for milk treated without altering its composition, or for milk, the fat content of which has been standardized under domestic legislation".

The entire commodity may be consumed.

<u>Portion of the commodity to which the MRL applies (and which is analyzed)</u>: *Whole commodity.*

The scientific species names of the relevant animals are not repeated in this group of commodities. For these names see Group 030 Meat (from mammals other than marine mammals).

Group 033   Milks

<u>Code No.</u>	<u>Commodity</u>
ML 106	**Milks**
ML 107	**Milk of cattle, goats and sheep**
ML 810	**Buffalo milk**
ML 811	**Camel milk**
ML 812	**Cattle milk**
ML 814	**Goat milk**
ML 822	**Sheep milk**

TYPE 7   Poultry products

Poultry meat (including Pigeon meat)

<u>Class B</u>
Type 7   Poultry products Group 036 Group Letter Code PM

Group 036. Poultry meats, are the muscular tissues including adhering fat and skin from poultry carcases as prepared for wholesale or retail distribution.

Exposure to pesticides may result from external treatment of animals or poultry houses or through animal metabolism following oral intake of pesticides with feed.

The entire product may be consumed.

<u>Portion of the commodity to which the MRL applies (and which is analyzed</u>: *Whole commodity (without bones): For fat-soluble pesticides a portion of adhering fat is analyzed and MRLs apply to the poultry fat.*

Group 036   Poultry meat (including Pigeon meat)

<u>Code No.</u>   <u>Commodity</u>

PM 110   **Poultry meat**

PM 840   **Chicken meat**
            Several breeds of *Gallus gallus* L., and other *Gallus* spp.

PM 841   **Duck meat**
            Breeds of *Anas platyrhynchos* L. and other *Anas* spp.

PM 842   **Goose meat**
            Anser anser L.; other *Anser* spp.

PM 843   **Guinea-fowl meat**
            Breeds of *Numida meleagris*

PM 844   **Partridge meat**
            Perdrix spp. and *Alectoris* spp.

PM 845   **Pheasant meat**
            Breeds of *Phasanius colchicus* and other *Phasanius* spp. and ssp.

PM 846 **Pigeon meat**
Breeds of *Columba livia* Gmelin; other *Columba* spp.;
Streptopelia spp.

PM 847 **Quail meat**
Coturnix coturnix (L.); *Colinus virginianus*; *Lophotyx californicus*;

PM 4831 **Quail, Bobwhite,** see Quail
Colinus virginianus

PM 4833 **Quail, California,** see Quail meat
Lephotyx californicus

PM 848 **Turkey meat**
Breeds of *Meleagris gallopavo* L.

Poultry fats

<u>Class B</u>

Type 7   Poultry products Group 037 Group Letter Code PF

Poultry fats are derived from the fatty tissues of poultry. Exposure to pesticides may result from external treatment of animals or poultry houses or through animal metabolism following oral intake of pesticides with feed.

The entire product may be consumed.

<u>Portion of the commodity to which the MRL applies (and which is analyzed)</u>: *Whole commodity.*

The scientific species names of the relevant animals are not repeated for this group of commodities. For these names see Group 036 Poultry meat.

Group 037   Poultry fats

<u>Code No.</u>   <u>Commodity</u>

PF 111   **Poultry fats**

PF 840   **Chicken fat**

PF 841   **Duck fat**

PF 842   **Goose fat**

PF 848   **Turkey fat**

Poultry, Edible offal of

<u>Class B</u>
Type 7   **Poultry products   Group 038   Group Letter Code P0**

Poultry edible offal are such edible tissues and organs, other than poultry meat and poultry fat, from slaughtered poultry as have been passed as fit for human consumption. Examples: liver, gizzard, heart, skin etc. In the former Classification of Food and Feed Groups in the Guide to Codex Maximum Limits for Pesticide Residues 1978: CAC/PR 1-1978 the name Poultry by-products was used for this group.

Exposure to pesticides is through animal metabolism following oral intake of pesticides with feed or may result from external treatment of animals or poultry houses.

The entire product may be consumed.

<u>Portion of the commodity to which the MRL applies (and which is analyzed)</u>: *Whole commodity.*

Group 038   **Poultry, Edible offal of**

<u>Code No.</u>	<u>Commodity</u>
PO 111	**Poultry, Edible offal of**
PO 113	**Poultry skin**
PO 840	**Chicken, Edible offal of**
PO 841	**Duck, Edible offal of**
PO 842	**Goose, Edible offal of**
PO 849	**Goose, liver**
PO 848	**Turkey, Edible offal of**

Eggs of Poultry

<u>Class B</u>
Type 7   Poultry products Group 039 Group Letter Code PE

Group 039. Eggs are the fresh edible portion of the body produced by female birds, especially domestic fowl.

The edible portion includes egg yolk and egg white after removal of the shell.

<u>Portion of the commodity to which the MRL applies (and which is analyzed)</u>: *Whole egg whites and yolks combined after removal of shell.*

The scientific species names of the relevant animals are not repeated for this group of commodities. For these names see Group 036 Poultry meats.

Group 039   Eggs

<u>Code No.</u>	<u>Commodity</u>
PE 112	**Eggs**
PE 840	**Chicken eggs**
PE 841	**Duck eggs**
PE 842	**Goose eggs**
PE 847	**Quail eggs**

Type 8   Aquatic animal products

Aquatic animal products are derived from the edible parts of various aquatic animals, usually wild, harvested for food.

Group 040-042 Fish

Group 040-042 fish are gilled, aquatic vertebrate and/or cartilaginous animals of various zoological families and species, usually wild, as caught and prepared for wholesale and retail distribution. Exposure to pesticides is through animal metabolism or through water pollution. The fleshy parts of the animals and, to a lesser extent, milt and roe are consumed.

Portion of the commodity to which the MRL applies (and which is analyzed): *Whole commodity (in general after removing the digestive tract).*

Freshwater fish

Class B

Type 8   Aquatic animal products Group 040 Group Letter Code WF

The freshwater fishes generally remain lifelong, including the spawning period, in fresh water (lakes, ponds, rivers and brooks).

Several species of freshwater fish are domesticated and bred in fish farms. Exposure of the latter species to pesticides is mainly through compounded fish feed and can also be through water pollution.

Group 040   Freshwater fish

Code No.	Commodity
WF 115	Freshwater fish
WF 4837	Amur pike, see Pike *Esox reicherti*
WF 855	Barbs *Puntius* spp.; syn: *Barbus Cuvier* spp.
WF 856	Black bass *Micropterus salmonides*; *Micropterus* spp.

WF 857   **Bluegill sunfish (or Bluegill bream)**
            Lepomis macrochirus

WF 858   **Bream**
            Abramis brama L.; other
            Abramis spp.

WF 4839   **Brown trout,** see Trout, Brown
            Group 041, Diadromous fishes

WF 859   **Carps**
            Cyprinus carpio L.;
            Ctenopharyngodon idella;
            other spp of the family *Cyprinidae*

WF 4841   **Carp, Common,** see Carps
            Cyprinus carpio L.

WF 4843   **Carp, Chinese,** see Carp, Grass

WF 4845   **Carp, Grass,** see Carps
            Ctenopharyngodon idella

WF 860   **Carp, Indian**
            Labeo rohita; *L. calbassa*;
            Catla catla; *Cirrhinus mrigala*

WF 861   **Catfishes (freshwater)**
            Ictalurus punctatus and other *Ictalurus* spp. (North America); *Pagrus* spp.
            (Africa); *Kryptopterus* spp. (Asia); *Clarias* spp. (Africa/Asia); *Silurus glanis* L.
            (Europe, USSR) *Tandanus tandanus* (Australia)

WF 4847   **Channel catfish,** see Catfishes (freshwater)
            Ictalurus punctatus

WF 869   **Cod, Murray**
            Maccullochella peeli (Australia)

WF 862   **Gobies, Freshwater**
            Gobio gobio L.;
            syn: *G. fluviatilus* Agass other freshwater spp. of the family *Gobiidae*

WF 863   **Gourami** (Asia)
Osphronemus goramy; *Trichogaster pectoralis*; Helostoma temmincki

WF 4849   **Northern pike,** see Pikes
Esox lucius L.

WF 4851   **Mozambique tilapia,** see Tilapia

WF 864   **Perch**
Perca fluviatilis L.; P. *flavescens*; *Aspledinotus grunniens*

WF 4853   **Perch, American yellow,** see Perch
Perca flavescens

WF 4855   **Perch, European,** see Perch
Perca fluviatilis L.

WF 870   **Perch, Golden**
Macquaria ambigua (Australia)

WF 4857   **Perch, White,** see Perch
Aspledinotus grunniens;
syn: *Pomoxis annularis* Raf.

WF 865   **Pike**
Esox lucius L.; E. *reicherti*

WF 866   **Pike-perch**
Stizostedium lucioperca L.;
syn: *Lucioperca sandra* Cuv.

WF 4859   **Rhinofishes,** see Carp, Indian
Labeo spp. among others *Labeo rohita*
Lubeo calbassa

WF 867   **Roaches**
Rutilus rutilus L.;
syn: *Leuciscus rutilus* L.; other
Rutilus (syn: *Leuciscus*) spp.

WF 868   **Tilapia**

Oreochromis mossambicus; syn: *Sarotherodon mossambicus*; *Tilapia mossambicus.*
Other *Oreochromis* (*Sarotherodon* or *Tilapia* species).

WF 4861   **White perch,** see Perch, white

WF 4863   **White crappie,** see Perch, white

Diadromous fish

Class B
Type 8   Aquatic animal products Group 041 Group Letter Code WD

The diadromous fishes in general migrate from the sea to brackish and/or freshwater and in the opposite direction.

The anadromous species spawn in fresh water (streams, small rivers and brooks) e.g. several salmon species, whereas eels spawn in the ocean. Some species, such as trout, are domesticated and do not migrate. They are bred in fish farms in ponds, mountain streams etc. The latter species especially may be exposed to pesticides through compounded fish feed and also through water pollution.

The fleshy parts of the animals and, to a lesser extent, roe and milt are consumed.

Portion of the commodity to which the MRL applies (and which is analyzed): *Whole commodity (in general after removing the digestive tract).*

Group 041   Diadromous fish

Code No.   Commodity

WD 120   **Diadromous fish**

WD 121   **Salmon, Pacific**
            according to Codex Stan. 36 and 37, 1981, includes *Oncorhynchus gorbuscha*;
            O. *keta*; O. kisutch; O. masou; O. *nerka*; O. tschawytscha

WD 123   **Trout**
            Salmo clarki, S. *gairdneri*: syn: S. *irrideus* Gibbons:
            S. *trutta* L.; syn: *Trutta trutta* L.; *Salvelinus* namaycush: S. *alpinus*;
            S. *salvelinus* L.

WD 4877   **Atlantic salmon,** see Salmon, Atlantic

WD 4867   **Arctic char,** see Trout
            Salvelinus alpinus

WD 898   **Barramundi**
            Lates calcarifer (Australia, Indo-Pacific)

WD 4869   **Brook trout,** see Trout
                Salvelinus sp.

WD 4871   **Brown trout,** see Trout
                Salmo trutta L.;
                syn: *Trutta trutta* L.

WD 4873   **Char,** see Lake trout

WD 4875   **Cherry salmon,** See Subgroup Salmon, Pacific
                Oncorhynchus masou

WD 4889   **Chinook salmon,** see Subgroup Salmon, Pacific
                Oncorhynchus tschawytscha

WD 4891   **Chum salmon,** see Subgroup Salmon, Pacific
                Oncorhynchus keta

WD 4893   **Coho salmon,** see Subgroup Salmon, Pacific
                Oncorhynchus kisutch

WD 4895   **Cutthroat trout,** see Trout
                Salmo clarki

WD 890   **Eels**
                Anguilla anguilla (L.); *A. japonica*; *A. rostrata*, *A. australis*
                A. reinhardtii

WD 4897   **Eel, American,** see Eels
                Anguilla rostrata

WD 4899   **Eel, Australian,** see Eels
                Anguilla australis
                A. reinhardtii

WD 4901   **Eel, European,** see Eels
                Anguilla anguilla L.

WD 4903   **Eel, Japanese,** see Eels
                Anguilla japonica

WD 4905   **Giant sea perch,** see Barramundi

WD 4907   **German trout,** see Trout
            Salmo trutta L.;
            syn: *Trutta trutta* L.

WD 4909   **Keta salmon,** see Subgroup Salmon, Pacific
            Oncorhynchus keta

WD 4911   **King salmon,** see Subgroup Salmon, Pacific
            Oncorhynchus tschawytscha

WD 4913   **Lake trout,** see Trout
            Savelinus namaycush

WD 4915   **Medium red salmon,** see Subgroup Salmon, Pacific
            Oncorhynchus kisutch

WD 891   **Milkfish**
            Chanos chanos

WD 897   **Nile perch**
            Lates niloticus

WD 4917   **Pacific salmon,** see Subgroup Salmon, Pacific

WD 892   **Paddle fish**
            Polyodon spathula; other species of the family *Polyodonthidae*

WD 4919   **Pink salmon,** see Subgroup Salmon, Pacific
            Oncorhynchus gorbuscha

WD 4921   **Rainbow trout,** see Trout
            Salmo gairdneri
            syn: S. *irrideus* Gibbons

WD 893   **Salmon, Atlantic,** see Atlantic salmon
            Salmo salar L.;
            syn: *Trutta salar* L.

WD 4923   **Salmon, Pacific,** see Subgroup Salmon, Pacific at the beginning of this group

WD 4925   **Sea catfish,** see Group 042: Marine fishes

WD 894   **Shad**
            Alosa spp.;
            Hilsa spp.

WD 4927   **Silver salmon,** see Subgroup Salmon, Pacific
            Oncnrhynchus kisutch

WD 895   **Smelt**
            Osmerus eperlanus L.;
            O. *mordax*; other *Osmerus* spp.

WD 4929   **Smelt, European,** see Smelt
            Osmerus eperlanus L.

WD 4931   **Smelt, Rainbow,** see Smelt
            Osmerus mordax

WD 4933   **Sockeye salmon,** see Subgroup Salmon, Pacific
            Oncorhynchus nerka

WD 4935   **Spring salmon,** see Subgroup Salmon, Pacific
            Oncorhynchus tschawytscha

WD 896   **Sturgeon**
            Acipencer sturio L.; other spp. of the family *Acipenceridae*

Marine fish

<u>Class B</u>
Type 8   **Aquatic animal products Group 042 Group Letter Code WS**

The marine fishes generally live in open seas. They are all or nearly all wild species, caught and prepared (often deep-frozen) for wholesale and retail distribution. Exposure to pesticides is mainly through water pollution and animal metabolism.

Especially the fleshy parts of the animals and to a lesser extent roe and milt are consumed.

<u>Portion of the commodity to which the MRL applies (and which is analyzed)</u>: *Whole commodity (in general, after removing the digestive tract).*

Group 042   Marine fish

<u>Code No.</u>   <u>Commodity</u>

WS 125   **Marine fish**

WS 126   **Cod and Cod-like fishes**
Cods, Haddocks, Hakes, Pollacks, Whiting.
For scientific names see individual species

WS 127   **Flat-fishes**
Brill, Dab, Flounders, Halibut, Plaice, Sole, Turbot.
For scientific names see individual species

WS 128   **Mackerel and Mackerel-like Fishes**
Mackerel; Horse and Jack Mackerels; King and Spanish Mackerel
For scientific names see individual species; see also next Sub-Group WS 0129

WS 129   **Mackerel and Jack Mackerel**
According to Codex Stan. 119-1981 includes species of the following families and genera Mackerel:
Scombridae:
Scomber spp.:
Rastrelliger spp.
Jack Mackerel: *Carangidae*:
Trachurus spp.; *Decapterus* spp.

WS 130   **Sardines and Sardine-type fishes**
             According to Codex Stan. 94-1981 small fishes of the following species:
             Sardina pilchardus (Walbaum) (European Sardine);
             Sardinops melanosticta;
             S. *neopilchardus*; S. *ocellata*;
             S. *sagax*; S. *caerulea* (Pilchards):
             Sardinella aurita Valanciennes; syn: S. *anchovia*;
             S. *brasiliensis*; S. *maderensis*
             (Sardinella's or oil-sardines):
             Clupea harengus L. (Atlantic herring, small fishes):
             Clupea antipodum, C. *bassensus*; C. *fuengensis*;
             Sprattus sprattus phalericus (Risso),
             syn: *Clupea sprattus* L. (Sprat);
             Hyperlophus vittatus
             Nematolosa vlaminghi;
             Etrumeus nicrops;
             Ethmidium maculatus;
             Engraulis anchoita (Argentine anchovy):
             E. *ringens* (Peruvian anchovy)

WS 131   **Sharks**
             Porbeagle, Requiem sharks, Smooth hounds, Spiny dogfish, Liveroil sharks
             For scientific names see individual species

WS 132   **Tuna and Bonito**
             According to Codex Stan. 70-1981 includes:
             Tunas: *Thunnus alalunga* (Bonnaterre):
             Th. *albacares*; Th. *atlanticus*;
             Th. *obesus*: Th. *thynnus maccoyii*;
             Th. *thynnus orientalis*;
             Th. *thynnus thunnus* L. Th. tongoll;
             Euthynnus affinus;
             Eu. *alletteratus*: Eu. *lineatus*:
             Eu. *pelamis* L.; syn: Katsuwonus *pelumis* L.
             Bonitos: *Sarda chilensis*: S. *orientalis*;
             S. *sarda* Bloch; S. velox

WS 4937   **Albacore,** see Subgroup Tuna and Bonito
             Thunnus alalunga (Bonnaterre)

WS 920 **Anchovies**
Engraulis encrasicolus (L.);
E. *japonicus*; E. mordax; E. *capensis*
For Argentine anchovy and Peruvian anchovy see Sub-Group
Sardines and Sardine-type fishes

WS 921 **Barracudas**
Sphyraena spp.

WS 4939 **Bigeye tuna,** see Tuna, Bigeye

WS 4940 **Blackfin tuna,** see Tuna, Blackfin

WS 922 **Bluefish**
Pomatomus saltatrix

WS 923 **Bogue**
Boops boops (L.); syn: Box boops Bonaparte

WS 924 **Bonito**
Sarda sarda (Bloch); S. *chiliensis*; S. *orientalis*; S. *velox*
see also Subgroup Tuna and Bonito

WS 4941 **Bonito, Atlantic,** see Bonito
Sarda sarda (Bloch)

WS 4943 **Bonito, Eastern Pacific,** see Bonito
Sarda chiliensis

WS 956 **Bream, Silver**
Acanthopagrus australis (Australia)

WS 4945 **Brill,** see Turbot
Scophthalmus rhombus (L.);
syn: *Rhombus laevis* Rondelet

WS 925 **Butterfish**
Species of the family *Stromateidae*

WS 926 **Capelin**
Mallotus villosus Müller

WS 4947 **Catfish, Sea,** see Wolffish

WS 4949   **Coalfish,** see Pollack
             Pollachius virens L.;
             syn: *Gadus virens* L.
             see also Subgroup Cod and Cod-like fishes

WS 927   **Cod**
             According to Codex Stan. 50-1981 includes
             Gadus morhua L.; syn: C. *collarius* L.; G. ogac Richardson;
             G. *macrocephalus* see also Subgroup Cod and Cod-like fishes

WS 4951   **Cod, Atlantic,** see Cod
             Gadus morhua L.;
             syn: G. *callarius* L.

WS 4953   **Cod, Greenland,** see Cod
             Gadus ogac Richardson

WS 4955   **Cod, Pacific,** see Cod
             Gadus macrocephalus

WS 928   **Conger or Conger eel**
             Conger conger (L.); C. *oceanicus*; C. *orbignyanus*; *Astroconger myriaster*

WS 4957   **Conger, European,** see Conger
             Conger conger (L.);
             syn: C. *vulgaris* Cuv.

WS 929   **Dab or Common dab**
             Limanda limanda L.
             see also Subgroup Flat-fishes

WS 930   **Dolphinfish**
             Coryphaena hippurus L.

WS 4959   **Dorado,** see Dolphinfish

WS 931   **Drums**
             Species of the family *Sciaenidae*

WS 4961   **European sardine,** see Subgroup Sardines and Sardine fishes
             Sardina pilchardus (Walbaum)

WS 932   **Flounders**
Platichthys flesus (L.);
syn: *Pleuronectus flesus* L.:
Atheresthes evermanni; A. *stomias*;
Glyptocephalus cynoglossus L.;
Limanda ferruginea
see also Subgroup Flat-fishes

WS 933   **Garfish**
Belone belone (L.);
syn: B. *acus Risso*

WS 934   **Haddock**
Gadus aeglefinus L.;
syn: *Melanngrammus aeglefinus* L.
see also Subgroup Cod and Cod-like fishes

WS 935   **Hakes**
Merlucius merlucius (L.);
other *Merlucius* spp.
see also Subgroup Cod and Cod-like fishes

WS 936   **Halibut**
Hippoglossus hippoglossus L.;
H. *stenolepis*;
Reinhardtius hippoglossoides Walbaum
see also Subgroup Flat-fishes

WS 4963   **Halibut, Atlantic,** see Halibut
Hippoglossus hippoglossus L.

WS 4965   **Halibut, Greenland,** see Halibut
Reinhardtius hippoglossoides Walbaum

WS 4967   **Halibut, Pacific,** see Halibut
Hippoglossus stenolepis

WS 937   **Herring**
Clupea harengus L.; C. *pallasi*; other *Clupea* spp.
N.B. For small fishes of these species see also Sub-Group Sardines and
Sardine-type fishes

WS 4969   **Herring, Atlantic,** see Herring
                    Clupea harengus L.

WS 4971   **Herring, Pacific,** see Herring
                    Clupea pallasi

WS 4973   **Horse mackerel,** see Jack Mackerel
                    Trachurus spp.; several spp.
                    see also Subgroup Mackerel and Jack Mackerel

WS 4975   **Indian mackerel,** see Mackerel
                    Rastrelliger *kanagurta*; other Rastrelliger spp.

WS 938   **Jack mackerel**
                    Trachurus spp.;
                    Decapterus spp.
                    see also Subgroup Mackerel and Jack Mackerel

WS 939   **King mackerel**
                    Scomberomorus spp., including S. *cavalla*; S. *comerson* S. *guttatus*
                    see also Subgroup Mackerel and Mackerel-like fishes

WS 940   **Ling**
                    Molva molva L.; M. byrkelange Walbaum;
                    syn: M. *dipterygia*; M. *elongata* Otto

WS 4977   **Liveroil shark,** see Subgroup Sharks
                    Galeorhinus galeus L.; other *Galeorhinus* spp.

WS 4979   **Longtail tuna,** see Tuna, Longtail

WS 941   **Mackerel**
                    Scomber scombrus L.; *Scomber japonicus*; other *Scomber* spp., Rastrelliger
                    kanagurta; R. *brachysoma*; other Rastrelliger spp.
                    see also Subgroup
                    (a) Mackerel and Jack Mackerel
                    (b) Mackerel and Mackerel-like fishes

WS 4981   **Mackerel, Atlantic,** see Mackerel
                    Scomber Scombrus L.

WS 4983   **Mackerel, Chub,** see Mackerel
                    Scomber japonicus

WS 4985   **Mackerel, Indian,** see Mackerel and Indian Mackerel
       Rastrelliger *kanagurta*

WS 4987   **Mackerel, Short,** see Mackerel
       Rastrelliger *brachysoma*

WS 942   **Menhaden**
       Brevoortia spp.

WS 943   **Mullets**
       (among others Mullet, red; Mullet, striped)
       Mugil cephalus
       Mullus surmuletus L.;
       other spp of the family *Mugilidae*

WS 4989   **Northern bluefin tuna,** see Subgroup Tuna and Bonito
       Thunnus thynnus thynnus L.

WS 944   **Ocean Perch**
       According to Codex Stan. 51-1981 includes
       Sebastus marinus L.; S. *mentella*; S. *viviparus* Kroyer,
       S. *alutus*; *Scorpaena dactyloptera* Delaroche; *Helicolenus maculatus*

WS 4991   **Oil sardine,** see Subgroup Sardines and Sardine type fishes
       Sardinella spp.

WS 945   **Plaice**
       Pleuronectus platessa L.;
       P. *quadrituberculata*
       see also Subgroup Flatfishes

WS 4993   **Plaice, Alaska,** see Plaice
       Pleuronectus quadrituberculata

WS 4995   **Plaice, European,** see Plaice
       Pleuronectus platessa L.

WS 946   **Pollack**
       Polachius polachius L.;
       syn: *Gadus polachius* L.
       see also Subgroup Cod and Cod-like fishes

WS 947   **Pomfret, Atlantic**
            Brama brama:
            syn: B. *raii* Bloch

WS 4997   **Porbeagle,** see Subgroup Sharks
            Lamna naaus (Bonaterre)

WS 948   **Rays**
            spp. of the family *Rajidae*

WS 4999   **Requiem shark,** see Subgroup Sharks
            spp. of the family *Carcharinidae* of the order of *Selachii*

WS 5001   **Salema,** see Boque
            Boops *salpa* (L.);
            syn: *Sarpa salpa* L.

WS 957   **Salmon,** Threadfin
            Polydactylus sheridani (Australia)

WS 5003   **Sardinella or Oil sardine**
            see Subgroup Sardines and Sardine-type fishes
            Sardinella spp.

WS 5005   **Sardine, European**
            see Subgroup Sardines and Sardine-type fishes
            Sardina pilchardus Walbaum

WS 5007   **Scad, see Jack Mackerel**
            Decapterus spp.

WS 5009   **Scorpion fishes,** see Ocean Perch
            Scorpaena dactyloptera Delaroche,
            syn: *Helicolenus dactylopterus* (Delaroche)
            other spp. of the family *Scorpaenidae*

WS 949   **Sea bass**
            Morone labrax L; syn: *Dicentrarchus tabrax* (L.); M. *saxatilus*; other Morone
            spp.

WS 950   **Sea bream**
            Pagellus centrodontus (Delaroche);
            P. erythrinus (L.) other *Pagellus* spp.

WS 5011   **Sea catfish,** see Wolffish
                   Anarichas spp.

WS 5013   **Seerfish,** see Spanish Mackerel and King Mackerel
                   Scomberomorus spp.

WS 5015   **Shark,** see Subgroup Sharks

WS 5017   **Skipjack tuna,** see Subgroup Tuna and Bonito
                   Katsuwonus pelamis L.;
                   syn: *Euthynnus pelamis* L.

WS 5019   **Smooth hounds,** see Subgroup Sharks
                   Mustelus spp.

WS 951   **Sole**
                   Solea solea L.;
                   syn: S. *vulgaris* Quensel
                   see also Subgroup Flat-fishes

WS 5021   **Southern bluefin tuna,** see Tuna, Bluefin
                   Thunnus maccoyii;
                   syn: *Thunnus thunnus maccoyii*

WS 5023   **Spanish mackerel,** see King mackerel
                   Scomberomorus spp.; among others S. *maculatus*; Sc. *tritor*; Sc. *niphonius*

WS 5025   **Spiny dogfish,** see Subgroup Sharks
                   Squalis acanthias L.; other *Squalis* spp.

WS 5027   **Tailor** (Australia), see Bluefish

WS 952   **Tuna,** see also Subgroup Tuna and Bonito
                   Thunnus spp.

WS 5029   **Tuna, Bigeye,** see Subgroup Tuna and Bonito
                   Thunnus obesus

WS 5031   **Tuna, Blackfin,** see Subgroup Tuna and Bonito
                   Thunnus atlanticus

WS 5033   **Tuna, Bluefin,** see Subgroup Tuna and Bonito
                   Thunnus thynnus L.; Th. *maccoyii*

WS 5035   **Tuna, Longtail,** see Subgroup Tuna and Bonito
Thunnus Tongoll

WS 5037   **Tuna, Skipjack,** see Subgroup Tuna and Bonito
Katsuwonis pelamis L.;
syn: *Euthynnus pelamis* L.

WS 5039   **Tuna, Yellowfin,** see Subgroup Tuna and Bonito
Thunnus albacares

WS 953   **Turbot**
Scophthalmus maximus L.;
syn: *Rhombus maximus* (L.)
see also Subgroup Flat-fishes

WS 5041   **Witch flounder,** see Flounders
Glyptocephalus cynoglossus L.

WS 953   **Whiting**
Gadus merlangus L.
see also Subgroup Cod and Cod-like fishes

WS 955   **Wolffish**
Anarhichas lupus L.;
A. *minor* Olafsson

WS 5043   **Yellowfin tuna,** see Tuna, Yellowfin

WS 5044   **Yellowtail flounder,** see Flounders
Limanda ferruginea

Fish roe (including milt = soft roe) and edible offal of fish

<u>Class B</u>
**Type 8   Aquatic animal products Group 043 Group Letter Code WR for the Roe
              WL for liver and other edible fish offal**

Fish roes are the edible reproductive bodies of several species of fish. Of some of these only the "hard roe", the female reproductive body, is used whereas both the "hard" and "soft" roe (milt) of other species is marketed.

The term roe used in the commodity description includes if relevant both types of roe.

The liver of some species is used as such for human consumption or for production of liver oils (e.g. cod liver oil).

Exposure to pesticides is through animal metabolism.

<u>Portion of the commodity to which the MRL applies (and which is analyzed)</u>: *Whole commodity.*

See for scientific species or family names and subgroup description Group 040-042. The relevant group is indicated after the commodity name with (f) freshwater fishes, (d) diadromous fishes and (m) marine fishes.

Group 043   Fish roe (including milt = soft roe) and edible offal of fish

<u>Code No.</u>	<u>Commodity</u>
WR 140	**Fish roe**
WR 121	**Salmon roe, Pacific (d)**
WL 131	**Shark liver (m)**
WR 922	**Bluefish roe (m)**
WR 927	**Cod roe (m)**
WR 930	**Dolphinfish roe (m)**
WR 932	**Flounder roe (m)**
WR 937	**Herring roe (m)**

WR 941   **Mackerel roe (m)**

WR 943   **Mullet roe (m)**

WR 893   **Salmon roe, Atlantic (d)**

WR 894   **Shad roe (d)**

WR 896   **Sturgeon roe (d)**

WL 927   **Cod liver (m)**

Marine mammals

<u>Class B</u>
Type 8   **Aquatic animal products Group 044 Group letter Code WM**

Several sea mammals are caught on a large scale. The meat of various species is used as food or feed in some areas of the world. The blubber (=whale or seal fat) fat and train oil (oil derived from whale fat) is used after processing as raw material in food or feed manufacture; the sperm oil, as well as the spermaceti (a waxy substance from the head of sperm whales) is mainly used in cosmetics and some other industrial products.

Exposure to pesticides is by consumption of contaminated prey or through water pollution.

The entire commodity except the bones and other inedible parts may be consumed.

<u>Portion of the commodity to which the MRL applies (and which is analyzed)</u>: *Whole commodity as marketed, without bones. For fat-soluble pesticides a portion of the fat is analyzed and MRLs apply to the fat.*

Group 044   Marine mammals

<u>Code No.</u>	<u>Commodity</u>
WM 141	**Marine mammals**
WM 142	**Fat of Dolphins, Seals and Whales, unprocessed**
WM 970	**Dolphins** spp. of the family *Dolphinidae*
WM 5045	**Dolphin, Bottlenose,** see Dolphins *Tursiops truncatus* (Mont.)
WM 5047	**Dolphin, Humpback,** see Dolphins
WM 5049	**Dolphin, Spinner,** see Dolphins *Stenella longirostris*
WM 5051	**Porpoise,** see Whales *Phocaena phocaena*

WM 5053   **Sea-lions,** see Seals
           Otaria spp.; *Eumetopius* spp.:
           Zalophus spp. (all Pacific Ocean)

WM 971   **Seals**
           spp. of the families *Otariidae, Phocidae* and *Trichechidae,*
           syn: *Odobenidae*

WM 5055   **Seal, Common,** see Seals
           Phoca vitulina

WM 5057   **Seals, Eared,** see Seals
           Otariidae spp.

WM 5059   **Seals, Earless,** see Seals
           Phocidae spp.

WM 5061   **Seals, Fur,** see Seals
           Arctocephalus pusillus (South Africa)
           A. *australis* (South America, Australia)
           Callorhinus ursinus (North Atlantic)

WM 5063   **Seal, Grey,** see Seals
           Halichoerus grypus (North Atlantic)

WM 5065   **Seal, Harp,** see Seals
           Pagophillus groenlandicus (North Atlantic)

WM 5067   **Seal, Hooded,** see Seals
           Cystophora cristata (North Atlantic)

WM 5069   **Seal, Ringed,** see Seals
           Phoca hispida (North Atlantic)

WM 972   **Whales**
           spp. of the zoological order of the *Cetacae*

WM 5071   **Whales, Baleen,** see Whales
           spp. of the family *Balaenopteridae*
           (Sub-order *Mystacoceti*)

WM 5073   **Whale, Blue,** see Whales
           Balaenoptera musculus

WM 5075   **Whale, False killer,** see Whales
             Pseudorca crassidens

WM 5077   **Whale, Fin,** see Whales
             Balaenoptera physalus

WM 5079   **Whale, Humpback,** see Whales
             Megaptera novacangliae

WM 5081   **Whale, Killer,** see Whales
             Orcinus orca

WM 5083   **Whale, Minke,** see Whales
             Balaenoptera acutorostrata

WM 5085   **Whale, Sei,** see Whales
             Balaenoptera borealis

WM 5087   **Whale, Short-finned pilot,** see Whales
             Globicephala macrorhynchus

WM 5089   **Whale, Sperm,** see Whales
             Physeter catodon

WM 5091   **Whales, Toothed,** see Whales
             spp. of the families *Physeteridae, Ziphiidae* and *Orcinus orca* (family *Delphinidae*)

Crustaceans

<u>Class B</u>
Type 8   Aquatic animal products Group 045 Group Letter Code WC

Crustaceans are aquatic animals of various species, wild or cultivated, which have an inedible chitinous outer shell.

A small number of species live in fresh water, but most species live in brackish water and/or in the sea.

Exposure to pesticides is through animal metabolism or water pollution.

Crustaceans are prepared for wholesale or retail distribution at a "raw" stage, often still alive, "raw" and deep-frozen, or cooked directly after catching and deep-frozen. Shrimps or prawns may also be parboiled and thereafter deep-frozen.

Although the cooked or parboiled crustaceans should be regarded as processed foods, the animals of this group are primarily classified in the Chapter on Primary food commodities, Type 8: Aquatic animal products, since several crustaceans are also marketed in a "raw" form, i.e. not exposed to temperatures sufficiently high to coagulate the protein at the surface. A short reference to processed Crustaceans is given in Type 17: Derived edible products of animal origin, Group 084 Crustaceans, processed.

The entire commodity except the shell may be consumed: the "raw" commodities, in general, after cooking.

<u>Portion of the commodity to which the MRL applies (and which is analyzed)</u>: *Whole commodity (especially with the small sized species) or the meat without the outer shell, be prepared for wholesale and retail distribution.*

Group 045   Crustaceans

<u>Code No.</u>	<u>Commodity</u>
WC 143	**Crustaceans**
WC 144	**Freshwater crustaceans**

WC 144   **Freshwater crustaceans**
                    Astacus spp. (Europe),
                    Procambarus spp. (USA);
                    Macrobrachium spp. (Asia, Australia, South and Middle America); species of
                    the family *Palaemonidae*

WC 145 **Marine crustaceans**
All species mentioned in this group, except those recorded as Freshwater Crustaceans

WC 146 **Crabs**
According to Codex Stan. 90-1981
edible species of the sub-order *Brachyura* of the order of the *Decapoda* and the species of the family *Lithodidae* (= King Crabs) *Scylla* spp. (Mud Crabs)

WC 976 **Freshwater crayfishes**
Astacus spp. (Europe);
Procambarus spp. (USA);
Eustacus spp. (Australia)

WC 977 **Freshwater shrimps or prawns,** (see Note 2)
Palaemon spp.; *Macrobrachium* spp.; *Cherax* spp.

WC 5093 **Langouste,** see Spiny Lobster

WC 978 **Lobsters**
According to Codex Stan. 95-1981 include
Homarus spp., family of *Nephropsidae* and spp. of the families *Palinuridae* and *Scyllaridae*, i.e., Spiny lobsters and Slipper lobsters

WC 5095 **Lobster, American,** see Lobsters
Homarus americanus

WC 5097 **Lobster, European,** see Lobsters
Homarus gammarus L.; syn: *Cancer gammarus* L.

WC 5099 **Lobster, Norway,** see Lobsters
Nephrops norvegicus L.;
syn: *Cancer norvegicus* L. (See also Note 1)

WC 5101 **Prawns,** see Shrimps or Prawns

WC 5103 **Prawn, Banana,** see Shrimps or Prawns
Penaeus merguiensis (Australia, Indo-Pacific)

WC 5105 **Prawn, Brown tiger,** see Shrimps or Prawns
Penaeus esculentus (Australia)

WC 5107   **Prawn, Caramote,** see Shrimps or Prawns
               Penaeus kerathurus Forskal (Mediterranean)

WC 5109   **Prawn, Common,** see Shrimps or Prawns
               Palaemon serratus Pennant (Europe, Mediterranean)

WC 5111   **Prawn, Eastern king,** see Shrimps or Prawns
               Penaeus plebejus (Australia, Indo-Pacific)

WC 5113   **Prawn, Endeavour,** see Shrimps or Prawns
               Penaeus endeavouri (Australia)

WC 5115   **Prawn, Giant tiger,** see Shrimps or Prawns
               Penaeus monodon (Australia, Indo-Pacific)

WC 5117   **Prawn, Green tiger,** see Shrimps or Prawns
               Penaeus semisulcatus (Indo-Pacific)

WC 5119   **Prawn, Japanese king,** see Shrimps or Prawns
               Penaeus japonicus (Asia)

WC 5121   **Prawn, Kuruma,** see Prawn, Japanese King

WC 5123   **Prawn, Northern,** see Shrimps or Prawns
               Penaeus borealis (Northern Atlantic)

WC 5125   **Prawn, Western king,** see Shrimps or Prawns
               Penaeus latisulcatus (Australia, Indo-Pacific)

WC 5127   **Rock lobster,** see Lobsters
               Jasus spp. (family *Palinuridae*)

WC 979   **Shrimps or Prawns,** (See Note 2)
               According to Codex Stan. 37-1981 and 92-1981 include spp. of the Families
               Crangonidae, Palaemonidae, (See Note 3), *Pandalidae* and *Penaidae*

WC 5129   **Shrimps, Common,** see Shrimps or Prawns
               Crangon crangon L.;
               syn: C. *vulgaris* Fabr. (Europe, Mediterranean)

WC 5131   **Shrimp, Deepwater rose,** see Shrimps or Prawns
               Parapenaeus longirostris Lucas (Atlantic)

WC 5133   **Shrimp, Northern brown,** see Shrimps or Prawns
                Penaeus aztecus (USA)

WC 5135   **Shrimp, Northern pink,** see Shrimps or Prawns
                Penaeus notialis;
                syn: P. *duorarum* (USA, West Africa)

WC 5137   **Shrimp, Northern white,** see Shrimps or Prawns
                Penaeus sertiferus (USA)

WC 5139   **Slipper lobster,** see Lobsters
                spp. of the Family *Scyllaridae*

WC 5141   **Spiny lobster,** see Lobsters
                Palinurus vulgaris Latreille, other *Palinurus* spp.

<u>Note 1</u>: In some countries, species such as the Norway lobster (*Nephrops norvegicus* L.) are included in the commodity "Prawns" with some qualifying designation, such as Dublin Bay Prawn or Prawn of Bantry Bay (both Ireland). The Codex Stan. 92-1981 on Quick Frozen Shrimps and Prawns does not prevent this practice, provided that the designation on the label ensures that the consumer will not be misled.

<u>Note 2</u>: There is no clear-cut distinction between Shrimps and Prawns. In several countries the commodity name Shrimps is used for the small species whereas the slightly larger ones are called Prawns. However, a species marked in certain regions of the world as "Prawn" may be called in the local English language in other areas a shrimp and visa versa, e.g., *Pandalus borcalis* is called Northern prawn or Deepwater prawn in the United Kingdom and the same species is named Pink shrimp in Canada. In Australia only the name Prawn is used for animals included in this commodity.

<u>Note 3</u>: Not including the Freshwater species of the *Palaemonidae*.

Type 9   **Frogs, lizards, snakes and turtles**

<u>Class B</u>
Type 9   **Amphibians and reptiles Group 048 Group Letter Code AR**

Frog, lizard, snake and turtle products are the edible parts from various animal species of the zoological classes *Amphibia* and *Reptilia*, usually wild, harvested for food. Some frog species are cultivated in a few European and Asian countries and to a small extent in the USA and marketed in the form of deep-frozen frog legs. The wild species are marketed in the same manner.

A few turtle species are raised from eggs or hatchlings in some tropical countries, especially the Green Turtle.

Exposure to pesticides is through animal metabolism.

The entire product, except the bones and the bony or horny outer shell (turtles), may be consumed.

<u>Portion of the commodity to which the MRL applies (and which is analyzed)</u>: *Whole commodity as marketed without the outer shell.*

Group 048   Frogs, lizards, snakes and turtles

<u>Code No.</u>   <u>Commodity</u>

AR 148   **Frogs, lizards, snakes and turtles**

AR 149   **Reptiles**
            Lizards, snakes, turtles

AR 5143   **Bullfrog, see Frogs**
            Rana catesbeiana, *R. tigrina*

AR 5145   **BullFrog, Indian, see Frogs**
            Rana tigrina

AR 990   **Frogs**
            Rana spp.; especially *Rana catesbeiana*;
            R. *esculenta* L.; R. *dactyla* Lesson; R. *ridibunda* Pall.;
            R. *tigrina*
            other spp. of the family *Ranidae*

AR 5147   **Frog, Agile,** see Frogs
Rana dalmatina Bonap.

AR 5149   **Frog, Common,** see Frogs
Rana temporaria L.

AR 5151   **Frog, Edible,** see Frogs
Rana esculenta L.

AR 5153   **Frog, Marsh,** see Frogs
Rana ridibunda Pall.

AR 5155   **Frog, Pool,** see Frogs
Rana lessonae Camer

AR 991   **Lizards**
Species of the zoological order *Lacertilia*

AR 992   **Snakes**
Several spp. of the zoological order *Ophidia*

AR 993   **Turtles**
Species of the zoological order *Chelonia*

AR 5157   **Turtle, Green,** see Turtles
Chelone midas L.;
syn: Ch. *viridis* Schneid.

AR 5159   **Turtle, Hawksbill,** see Turtles
Eretmochelys imbricata

AR 5161   **Turtle, Loggerhead,** see Turtles
Caretta caretta L.;
syn: *Thalassochelys caretta* L.

Type 10   Molluscs (including Cephalopods) and other invertebrate animals

<u>Class B</u>
Type 10   Invertebrate animals Group 049 Group Letter Code IM

Molluscs are aquatic or land animals or various species, wild or cultivated, which have an inedible outer or inner shell.

The edible aquatic Molluscs live mainly in brackish water or in the sea; several species are cultivated. A few edible species of land snails are cultivated.

Exposure to pesticides is through animal metabolism: the aquatic species also through water contamination.

The entire commodity except the outer or inner shell may be consumed.

<u>Portion of the commodity to which the MRL applies (and which is analyzed)</u>: *Whole commodity after removal of shell.*

Group 049   Molluscs (including Cephalopods) and other invertebrate animals

<u>Code No.</u>   <u>Commodity</u>

IM 150   **Molluscs, including Cephalopods**

IM 151   **Marine bivalve molluscs**
                  Sub-class *Lamellibranchia*
                  Clams, Cockles, Mussels, Oysters, Scallops

IM 152   **Cephalopods**
                  Cuttlefishes, Octopuses, Squids

IM 5163   **Beche-de-mer,** see Sea-cucumbers

IM 1000   **Clams**
                  Species of the families *Arcidae*; *Mactridae*; *Veneridae*

IM 1001   **Cockles**
                  Cardium edule L.; other *Cardium* spp.

IM 5165   **Cockle, Common,** see Cockles
                  Cardium edule L.

IM 1002   **Cuttlefishes**
            Sepia officinalis L.; *S. elegans* d'Orbigny; other Sepia spp.; *Sepiola atlantica* d'Orbigny; S. *rondeleti* Leach

IM 5167   **Cuttlefish, Common,** see Cuttlefishes
            Sepia officinalis L.

IM 5169   **Giant snail,** see Snails, Edible (Africa, Asia)
            Achatina fulica fer.; *A. achatina*; *Archachatina* spp.

IM 1003   **Mussels**
            Mytilus edulis L. (Europe);
            M. *galloprovincialis* Lam. (Mediterranean):
            M. *smaragdinus* (Asia); other *Mytilus* spp.

IM 5171   **Little cuttle,** see Cuttlefishes
            Sepiola atlantica d'Orbigny; S. *rondeleti* Leach

IM 5173   **Octopuses**
            Octopus vulgaris Lam.;
            Eledone cirrhosa Lam.;
            E. *moschata* Lam.

IM 5175   **Octopus, Common,** see Octopuses
            Octopus vulgaris Lam.

IM 5177   **Octopus, Curled,** see Octopuses
            Eledone cirrhosa Lam.

IM 5179   **Octopus, Musky,** see Octopuses
            Eledone moschata Lam.

IM 1004   **Oysters** (including Cupped oysters)
            Ostrea edulis L.; other *Ostrea* spp.; *Crassostrea angulata* Lam.; syn: *Gryphaea angulata* Lam.; *Crassostrea gigas*; C. *virginica*; other *Crassostrea* spp.

IM 5181   **Oyster, American cupped,** see Oysters
            Crassostrea virginica (American)

IM 5183   **Oyster, European,** see Oysters
            Ostrea edulis L.

IM 5185   **Oyster, Pacific cupped,** see Oysters
           Crassostrea gigas (Asia, Canada)

IM 5187   **Oyster, Portuguese cupped,** see Oysters
           Crassostrea angulata Lam.;
           syn: *Gryphaea angulata* Lam. (S.W. Europe)

IM 5189   **Oyster, Sydney rock,** see Oysters (including Cupped Oysters)
           Crassostrea commercalis (Australia)

IM 1005   **Scallops**
           Pecten spp.; *Placopecten* spp.:
           Argopecten sp.

IM 5191   **Scallop, Australian,** see Scallops
           Pecten meridionalis (Australia)

IM 5193   **Scallop, Bay,** see Scallops
           Argopecten irradians (N. America)

IM 5195   **Scallop, Giant Pacific,** see Scallops
           Pecten caurinus (America)

IM 5197   **Scallop, Great,** see Scallops
           Pecten maximus (L.) (W. Europe, Mediterranean)

IM 5199   **Scallop, New Zealand,** see Scallops
           Pecten novaezealandiae (New Zealand)

IM 5201   **Scallop, Queen,** see Scallops
           Pecten opercularis (L.);
           syn: *Chlamys opercularis* L. (W. Europe)

IM 5203   **Scallop, Sea,** see Scallops
           Placopecten magellanicus (N. America)

IM 1010   **Sea-cucumbers**
           species of the zoological order of the *Holothuroidea*

IM 1006   **Sea urchins**
           Species of the zoological order of the *Echinoidea*

IM 1007   **Snails, Edible**
           Helix spp.; *Achatina* spp.

IM 5205   **Snail, Garden,** see Snails, Edible
           Helix aspersa Müller

IM 5207   **Snail, Giant,** see Snails, Edible
           Achatina fulica Fer.; *A. achatina*

IM 5209   **Snail, Roman,** see Snails, Edible
           Helix pomatia L.

IM 1008   **Squids**
           Loligo forbesi Steenstrup
           L. vulgaris Lam.; other *Loligo* spp.
           Allotheuthis subulata lam; *Ommastrephes sagittatus* Lam.;
           syn: *Todarodes sagittatus* Lam.,
           T. pacificus;
           Illex illecebrosus, other *Illex* spp.

IM 1009   **Squid, Common,** see Squids
           Loligo forbesi Steenstrup

IM 5211   **Squid, European flying,** see Squids
           Ommastrephes sagittatus Lam.;
           syn: *Todarodes sagittatus* Lam. (Europe)

IM 5213   **Squid, Japanese Flying,** see Squids
           Todarodes pacificus (Asia)

IM 5215   **Squid, Short-finned,** see Squids
           Illex illecebrosus

CLASS C   PRIMARY FEED COMMODITIES

For the purpose of the Codex Alimentarius the term "primary feed commodity" means the product in or nearly in its natural state intended for sale to:

(a)   the stock farmer as feed which is used without further processing for livestock animals or after silaging or similar farm processes;

(b)   the animal feed industry as a raw material for preparing compounded feeds;

Type 11   Primary feed commodities of plant origin

The primary feed commodities of plant origin include products after removal of certain parts of the plants.

Some types of the primary feed commodities are grown and used exclusively for animal feeding purposes, e.g. alfalfa, vetch and maize forage. Other types are derived from crops of which the edible parts are used directly or after processing as food, whereas the "waste" parts of these crops are generally used for feeding purposes, e.g. cereal straws, pea vines (fresh = green), pea hay, maize fodder, sugar beet tops or leaves.

Legume animal feeds

Class C
Type 11   Primary feed commodities of plant origin
Group 050 Group Letter Code AL

Group 050. Legume animal feeds include various species of leguminous plants used for animal forage, grazing, fodder hay or silage, with or without seed. Several species are grown exclusively for animal feeding purposes, whereas some others are grown primarily as food crops. The "waste" parts of the latter crops are often used as animal feed, either in the fresh form or as hay.

The entire commodity may be consumed by livestock animals.

Portion of the commodity to which the MRL applies (and which is analyzed): *Whole commodity as presented for wholesale or retail distribution.*

In view of the wide range of moisture contents in most animal feeds, except straw, moving in commerce, the MRLs should preferably be set end expressed on a "dry-weight" basis.

A "dry-weight" basis implies that the commodity is analyzed for pesticide residues as received, that the moisture content is determined, preferably by a standard method for use on the relevant commodity, and the residue content is then calculated as if it were wholly contained ln the dry matter (Ref. Report 1980 JMPR[1]).

The residues are expressed on a dry-weight basis if not otherwise stated. To avoid confusion caused by the not always consistent commodity description in the past, the "dry-weight" basis, will be indicated, if relevant, with the designation "dry weight" after the residue figure e.g.

pea vines (green); x mg/kg dry weight
pea hay   ; x mg/kg dry weight

Group 050   Legume animal feeds

<u>**Code No.**</u>   <u>**Commodity**</u>

AL 157   **Legume animal feeds**

AL 61   **Bean fodder**
                 Phaseolus spp.

AL 72   **Pea hay or Pea fodder (dry)**

AL 1020   **Alfalfa fodder**
                 Medicago sativa L., subsp. *sativa* L.:
                 M. *sativa* L., subsp. *falcata* (L.) Arcang, and hybrids = M. *sativa* L., subsp.
                 varia (Martijn) Arcang

AL 1021   **Alfalfa forage** (green)
                 for scientific names see AL 1020 Alfalfa fodder

AL 1030   **Bean forage (green)**

AL 1022   **Bean, Velvet**
                 Mucuna deeringiana (Bort.) Merr.;
                 syn: *Stizolobium deeringianum* Bort.;
                 other *Stizolobium* spp.

AL 5217   **Chickling vetch,** see Vetch, Chickling

AL 524   **Chick-pea fodder**
                 Cicer arietinum L.

---

[1]   FAO Plant Production and Protection Paper No. 26, Rome 1981.

AL 1023   **Clover**
           Trifolium, several spp. and ssp.
           Melilotus spp.

AL 1031   **Clover hay or fodder**

AL 5219   **Grass pea,** see Vetch, Chickling

AL 1024   **Kudzu**
           Pueraria lobata (Willd.) Ohwi;
           syn: P *thunbergiana* (Sieb. & Zucc.) Benth.
           P. *phaseoloides* (Roxb.) Benth.

AL 5221   **Kudzu, Tropical,** see Kudzu
           Pueraria phaseoloides (Roxb.) Benth.

AL 1025   **Lespedeza**
           Lespedeza cuneata (Dum.) G. Don;
           syn: L. *sericea* Miq.

AL 545   **Lupin,** forage
           among others *Lupinus albus* L.;
           L. *angustifolius* L.; L. *luteus* L.; sweet varieties

AL 5223   **Melilot,** see Clovers
           Melilotus spp.

AL 528   **Pea vines** (green)

AL 697   **Peanut fodder**
           Arachis hypogaea L.

AL 1270   **Peanut forage (green)**

AL 5227   **Puero,** see Kudzu, Tropical
           Pueraria phaseoloides (Roxb.) Benth.

AL 1027   **Sainfoin**
           Onobrychis viciifolia Scop.;
           syn: O. *sativa* Lamk.

AL 5229   **Sericea,** see Lespedeza

AL 541   **Soya bean fodder**
              Glycine max (L.) Merr; for synonyms see VP 0541

AL 1265   **Soya bean forage (green)**

AL 1028   **Trefoil**
              Lotus corniculatus, L.; other *Lotus* spp.

AL 5231   **Tropical kudzu,** see Kudzu, Tropical

AL 5233   **Velvet bean,** see Bean, Velvet

AL 1029   **Vetch**
              Vicia spp., several ssp.; *Astralagus* spp.;
              Coronilla varia L.; *Lathyrus sativus* L.

AL 5235   **Vetch, Chickling,** see Vetch
              Lathyrus sativus L.

AL 5237   **Vetch, Crown,** see Vetch
              Coronilla varia L.

AL 5239   **Vetch, Milk,** see Vetch
              Astralagus spp.

Straw, fodder and forage of cereal grains and grasses, except grasses for sugar production (including buckwheat fodder)

<u>Class C</u>
Type 11   Primary feed commodities of plant origin
　　　　　　　　　Group 051 Group Letter Codes AS (straws and fodders dry) AF (forage)

The straw, fodder and forage of cereal grains are derived from various plants of the grass family (*Gramineae*).

Cereal grains are grown to a limited extent as a forage crop. The immature crop is fed to livestock animals as succulent forage or as silage.

The cereal grain crops are mainly grown for human food or raw material for preparing food products. The "waste" parts remaining after harvest of the grain kernels (stems, stalks, leaves and empty ears) are extensively used and distributed for animal feeding purposes, in the form of dry fodder or straw.

Several other species of the grass family are exclusively grown as forage crops. These crops are either used for grazing or are prepared for wholesale or retail distribution in the form of grass silage (in general one or more cuttings from immature plants), as artificially dried grass or as hay.

The entire commodity may be consumed by livestock animals.

<u>Portion of the commodity to which the MRL applies (and which is analyzed)</u>: *Whole commodity, as presented for wholesale or retail distribution.*

In view of the wide range of moisture contents in the animal feeds of this group, except the straws and hays, moving in commerce the MRLs should preferably be set and expressed on a "dry-weight" basis.

A "dry-weight" basis implies that the commodity is analyzed for pesticide residues as received, that the moisture content is determined, preferably by a standard method for use on the relevant commodity, and that the residue content is then calculated as if it were wholly contained in the dry matter (Ref. Report 1980 JMPR).

The residues on the dry commodities of this group, e.g. straws and hays, are expressed on the commodity as such (see explanatory note below).

Group 051   **Straw, fodder and forage of cereal grains and grasses, except grasses for sugar production (including buckwheat fodder)**

<u>Code No.</u>   <u>Commodity</u>

AS 161   **Straw, fodder (dry) and hay of cereal grains and other grass-like plants**

AS 81   **Straw and fodder (dry) of cereal grains**

AS 162   **Hay or fodder (dry) of grasses**

AS 640   **Barley straw and fodder, dry**

AS 5241   **Bermuda grass**
                 Cynodon dactylon (L.) Pers.
                 see Subgroup Hay or Fodder (dry) of Grasses

AS 5243   **Bluegrass**
                 Poa spp.
                 see Subgroup Hay or Fodder (dry) of Grasses

AS 5245   **Brome grass**
                 Bromus spp.
                 see Subgroup Hay or Fodder (dry) of Grasses

AS 641   **Buckwheat fodder**
                 Fagopyrum esculentum Moench;
                 syn: F. *sagittatum* Gilib.

AS 5247   **Corn fodder,** see Maize fodder

AF 5249   **Corn forage,** see Maize forage

AS 5251   **Darnel**
                 Lolium spp.
                 see Subgroup Hay or Fodder (dry) of Grasses

AS 5253   **Fescue**
                 Festuca spp.
                 see Subgroup Hay or Fodder (dry) of Grasses

AF 645   **Maize forage**
                 Zea Mays L.

AS 645   **Maize fodder**
                    Zea Mays L.

AS 646   **Millet fodder,** dry
                    Echinochloa cruss-galli (L.) Beauv.; *Eleusine coracana* (L.) Gaertn.;
                    Panicum miliaceum L.; *Penisetum typhoides* (Burm. f.) Stapf & Hubhard;
                    Setaria italica (L.) Beauv.;
                    For synonyms see the specific Millets in Group 021: Grasses

AF 647   **Oat forage (green)**
                    Avena fatua L., A. *abysinnica* Hochst.

AS 647   **Oat straw and fodder,** dry
                    Avena fatua L. A. *abysinnica* Hochst.

AS 649   **Rice straw and fodder,** dry
                    Oryza sativa L.

AF 650   **Rye forage (green)**
                    Secale cereale L.

AS 650   **Rye straw and fodder,** dry

AF 651   **Sorghum forage (green)**
                    Sorghum bicolor (L.) Moench); other Sorghum spp.

AS 651   **Sorghum straw and fodder,** dry

AS 657   **Teosinte fodder**
                    Zea mays ssp. mexicana (Schrader) Iltis;
                    syn: *Z. mexicana* (Schrader) Kunze; *Euchleaena mexicana* Schrader

AS 654   **Wheat straw and fodder,** dry
                    Triticum aestivum L.; T. *vulgare* Vill.; other *Triticum* spp.

Explanatory Note: Another advantage of expressing the residue on a "dry-weight" basis is that it overcomes the problems arising from the often inconsistent use of the terms forage and fodder. **Forage:** Crops grown exclusively for animal feed. These crops are either used for grazing or are prepared as silage or as dry fodder. **Fodder:** Coarse feed for livestock animals, especially cattle, horses and sheep, such as straw, hay, maize, stalks (stover) etc. e.g. Maize forage: whole green plant, prior to maturity including the immature or nearly mature cobs). **Maize fodder:** stover or whole stalks (with ears removed) remaining after the harvest of the mature and sun-dried cobs.

Miscellaneous Fodder and Forage crops

<u>Class C</u>
Type 11   Primary feed commodities of plant origin
**   Group 052 Group Letter Codes AM (fodder) AV (forage)**

Group 052. Miscellaneous Fodder and Forage crops, are derived from various kinds of plants except leguminous and grassy plants (family *Gramineae*). However, for convenience, the fodders and forage of grasses for sugar production are included in this group. Some of the crops listed in this group are primarily grown for human food or as raw material for preparing food (e.g. sugar beet) and the "waste" material of such crops is used as animal feed.

The entire commodity may be consumed by livestock animals, either in a succulent form, as silage or in the form of dry fodder.

<u>Portion of the commodity to which the MRL applies (and which is analyzed)</u>: *Whole commodity as presented for wholesale or retail distribution. In view of the wide range of moisture contents in the animal feeds of this group moving in commerce the MRLs should, if relevant, preferably be set and expressed on a "dry-weight" basis, see explanation in Group O50 Legume animal feeds.*

Group 052   Miscellaneous Fodder and Forage crops

<u>Code No.</u>   <u>Commodity</u>

AM 165   **Miscellaneous fodder and forage crops**
            except leguminous and grassy plants (*Gramineae*), but including grasses for sugar production

AM 691   **Cotton fodder, dry**
            Gossypium spp.

AV 1050   **Cow cabbage**
            Brassica oleracea L., var. *acephala* (D.C.) Alef subvar *viridis*

AM 1051   **Fodder beet**
            Beta vulgaris var. *rapa*

AV 1051   **Fodder beet leaves or tops**

AM 5255   **Mangel or Mangold,** see Fodder beet

AM 5256   **Mangoldwurzel,** see Fodder beet

AV 1052   **Marrow-stem cabbage or Marrow-stem kale**
           Brassica oleracea L, convar, *acephala* (D.C.) Alef.
           var. *medullosa* Thell.

AM 738   **Mint hay**

AM 353   **Pineapple fodder**

AM 659   **Sugar cane fodder**
           Saccharum officinarum L.

AV 659   **Sugar cane forage**

AM 497   **Swedish turnip or Swede fodder**
           Brassica napus L., var. *napobrassica* (L.) Rchb.; syn: R. *napobrassica* (L.)
           Mill.

AM 506   **Turnip fodder**
           B. *campestris* L., ssp *rapifera* (Metzg) Sinsk;
           syn: B *rapa* L., var. *rapa*

AV 506   **Turnip leaves or tops**

CLASS D and E PROCESSED FOODS

The term "processed food" means the product, resulting from the application of physical, chemical or biological processes or combinations of these to a "primary food commodity", intended for direct sale to the consumer, for direct use as an ingredient in the manufacture of food or for further processing.

"Primary food commodities" treated with ionizing radiation, washed, sorted or submitted to similar treatment are not considered to be "processed foods".

<u>Class D</u>
Processed food of plant origin

<u>Class D</u>
Type 12   Secondary food commodities of plant origin

The term "secondary food commodity" means a "primary food commodity" which has undergone simple processing, such as removal of certain portions, drying (except natural drying), husking, and comminution, which do not basically alter the composition or identity of the product. Natural field dried mature crops or parts of crops such as pulses, bulb onions or cereal grains are not considered as secondary food commodities.

Secondary food commodities may be processed further or used as ingredients in the manufacture of food or sold directly to the consumer.

Dried fruits

<u>Class D</u>
Type 12   Secondary food commodities of plant origin
Group 055 Group Letter Code DF

Group 055. Dried fruits. The commodities of this group are in general artificially dried. They may or may not be preserved or candied with addition of sugars.

Exposure to pesticides may arise from pre-harvest applications, post-harvest treatment of the fruits before processing, or treatment of the dried fruit to avoid losses during transport and wholesale or retail distribution.

<u>Portion of the commodity to which the MRL applies (and which is analyzed)</u>: *Whole commodity after removal of stones, but the residue is calculated on the whole commodity.*

Group 055   Dried Fruits

Code No.   Commodity

DF 167   **Dried fruits**

DF 14   **Prunes**
Prunus domestica L.

DF 226   **Apples, dried**
Malus domesticus Borkhausen

DF 240   **Apricots, dried**
Prunus armeniaca L.;
syn: *Armeniaca vulgaris* Lamarck

DF 5257   **Currants**
1.   Seedless blue grape var., dried, see Dried grapes
Vitis vinifera L., var.
2.   See Currants, Black, Red, White, Group 4 Berries
and other small fruits

DF 269   **Dried grapes** (= Currants, Raisins and Sultanas)
Vitis vinifera L., var. *corinthiaca* and var. *apyrena*

DF 5259   **Dried vine fruits,** see Dried grapes

DF 295   **Dates, dried or dried and candied**
Phoenix dactylifera L.

DF 297   **Figs, dried or dried and candied**
Ficus carica L.

DF 5261   **Muscatel,** see Dried grapes

DF 5263   **Raisins** (seedless white grape var., partially dried), see Dried grapes
Vitis vinifera L.;

DF 5265   **Sultanas,** see Dried grapes

DRIED VEGETABLES

<u>Class D</u>
Type 12   Secondary food commodities of plant origin
**   Group 056 Group Letter Code DV**

Group 056. Dried vegetables. The commodities of this group are in general artificially dried and often comminuted. Exposure to pesticides is from pre-harvest applications and/or treatment of the dry commodities.

The entire commodity may be consumed after soaking or boiling.

<u>Portion of the commodity to which the MRL applies (and which is analyzed</u>: *Whole commodity as prepared for wholesale or retail distribution.*

Group 056   Dried vegetables

<u>Code No.</u>   <u>Commodity</u>

DV 168   **Dried Vegetables**

<u>Class D</u>
Type 12   Secondary food commodities of plant origin
**   Group 057 Group Letter Code DH**

Group D57. Dried herbs. The commodities of this group are in general artificially dried and often comminuted. For the commodities in the "fresh" state see Group 027 Herbs. Exposure to pesticides is from pre-harvest applications and/or treatment of the dry commodities.

They are consumed in the dried form or soaked as a condiment in food commodities of plant or animal origin or in drinks, generally in small amounts.

<u>Portion of the commodity to which the MRL applies (and which is analyzed)</u>: *Whole commodity as prepared for wholesale or retail distribution.*

Group 057   Dried herbs

<u>Code No.</u>   <u>Commodity</u>

DH 170   **Dried herbs**

DH 720   **Angelica, including Garden Angelica, dry**
             Angelica sylvestris L.; A. *archangelica* L.

DH 721   **Balm leaves, dry**
                   Melissa officinalis L.

DH 722   **Basil, dry**
                   Ocimum basilicum L.

DH   7   **Bay leaves, dry**
                   Laurus nobilis L.

DH 724   **Borage, dry**
                   Borago officinalis L.

DH 728   **Burning bush,** dry
                   Dictamnus albus L.;
                   syn: D. *fraxinella* Pers

DH 726   **Catmint, dry**
                   Nepeta cataria L.

DH 624   **Celery leaves, dry**
                   Apium graveolens L.

DH 5269   **Cretan Dittany, dry,** see Burning bush, dry

DH 731   **Fennel, dry**
                   Foeniculum vulgare Mill.;
                   syn: F *officinale* All.; F *capillaceum* Gilib.

DH 1100   **Hops, dry**
                   Humulus lupulus L.

DH 732   **Horehound, dry**
                   Marrubium vulgare L.

DH 733   **Hyssop, dry**
                   Hyssopus officinalis

DH 734   **Lavender, dry**
                   Lavendula angustifolia Mill.;
                   syn: L. *officinalis* Chaix; L. *spica* L.; L. *vera* DC.

DH 755   **Lovage, dry**
                   Levisticum officinale Koch.

DH 736   **Marjoram, dry**
          Marjorana hortensis Moench.;
          syn: *Origanum marjorana* L.; *Origanum vulgare* L.

DH 738   **Mints, dry**
          Several Mint species and hybrids and *Pulegium vulgare* Mill;
          (see also individual Mints, Group 027 Herbs)

DH 5271   **Oregano** (= Wild Marjoram) dry, see Marjoram
          Origanum vulgare L.

DH 741   **Rosemary, dry**
          Rosmarinus officinalis L.

DH 742   **Rue, dry**
          Ruta graveolens L.

DH 743   **Sage, dry**
          Salvia officinalis L.; *S. sclarea* L.

DH 745   **Savory, Summer; Winter, dry**
          Satureja hortensis L.; *S. montana* L.

DH 747   **Sweet Cicely, dry**
          Myrrhis odorata (L.) Scop.

DH 748   **Tansy and related species, dry**
          Tanacetum vulgare L.; *T. balsamita* L .;
          syn: *Chrysanthemum balsamita* L.

DH 750   **Thyme, dry**
          a.o. *Thymus vulgaris* L.; Th. *serpyllum* L. and *Thymus* hybrids

DH 752   **Wintergreen leaves, dry**
          Gaultheria procumbens L.;
          (not included herbs of the Wintergreen family, *Pyrolaceae*)

DH 753   **Woodruff, dry**
          Asperula odorata L.

DH 754   **Wormwoods, dry**
          Artemisia absinthium L.; *A. abrotanum* L.;
          A. vulgaris L.

Milled cereal products (early milling stages)

<u>Class D</u>

Type 12   Secondary food commodities of plant origin
**   Group 058 Group Letter Code CM**

For final milling fractions, whether processed or not, see Group 065 Cereal grain milling fractions

Group 058. Milled cereal products (early milling stages). The group includes the early milling fractions of cereal grains, except buckwheat, cañihua and quinoa such as husked rice, polished rice and the unprocessed cereal grain brans.

Exposure to pesticides is through pre-harvest treatments of the growing cereal grain crop and especially through post-harvest treatment of cereal grains.

The entire commodity may be consumed after further processing or household preparation.

<u>Portion of the commodity to which the MRL applies (and which is analyzed)</u>: *Whole commodity as prepared for wholesale or retail distribution.*

<u>Note</u>: In view of the number of related commodities in this group some extra code numbers had to be used, not related to the primary food commodity from which the processed commodity concerned is prepared. Reference to the additional code numbers is given with the primary food commodity where it is listed for the first time in the Classification.

Group 058   Milled cereal products (early milling stages)

<u>**Code No.**</u>   <u>**Commodity**</u>

CM   81   **Bran, unprocessed of cereal grain**
                  (except buckwheat, cañihua and quinoa)

CM 1206   **Rice bran, unprocessed**

CM 649   **Rice, husked**

CM 1205   **Rice, polished**

CM 650   **Rye bran, unprocessed**

CM 654   **Wheat bran, unprocessed**

Miscellaneous secondary food commodities of plant origin

<u>Class D</u>

Type 12   Secondary food commodities of plant origin
Group 059 Group Code SM

Group 059 Miscellaneous secondary food commodities of plant origin

<u>Portion of commodity to which the MRL applies (and which is analyzed)</u>: *Whole commodity.*

<u>Code No.</u>   <u>Commodity</u>

SM 716   **Coffee beans, roasted**

Type 13   Derived edible products of plant origin

"Derived edible products" are foods or edible substances isolated from primary food commodities or raw agricultural commodities, not intended for human consumption as such, using physical, biological or chemical processing.

This type of processed food includes groups such as vegetable oils (crude and refined), by-products of the fractionation of cereals, fruit juices, teas (fermented and dried), cocoa powder and by-products of cocoa manufacturing, and extracts of various plants.

Cereal grain milling fractions

Class D
Type 13   Derived products of plant origin
Group 065 Group Letter Code CF

Group 065. Cereal grain milling fractions includes milling fractions of cereal grains at the final stage of milling and separation in the fractions. The group also includes the processed brans, as prepared for direct consumption.

Portion of commodity to which the MRL applies (and which is analyzed): ***Whole commodity.***

Note: In view of the number of related commodities in this group some extra code numbers had to be used, not related to the primary food commodity from which the commodity concerned is prepared. Reference to the additional code numbers is given with the primary food commodity where it is listed for the first time in the Classification.

Group 065 Cereal grain milling fractions

Code No.	Commodity
CF 81	**Cereal brans, processed**
CF 5273	**Corn flour**, see Maize flour
CF 5275	**Corn meal**, see Maize meal
CF 1255	**Maize flour**
CF 645	**Maize meal**
CF 649	**Rice bran, processed**
CF 650	**Rye bran, processed**

CF 1250   **Rye flour**

CF 1251   **Rye wholemeal**

CF 654   **Wheat bran, processed**

CF 1210   **Wheat germ**

CF 1211   **Wheat flour**

CF 1212   **Wheat wholemeal**

Teas

<u>Class D</u>
Type 13   Derived edible products of plant origin
**   Group 066   Group Letter Code DT**

Teas, Group 066, are derived from the leaves of several plants, principally *Camellia sinensis*.

They are used mainly in a fermented and dried form or only as dry leaves for the preparation of infusions, which are used as beverages.

Newly grown vegetative shoots (terminal bud and 2-3 leaves) of tea are plucked, withered, twisted and comminuted and thereafter, in general, fermented and dried.

Teas made from other plants are often prepared in a similar way.

<u>Portion of commodity to which the MRL applies (and which is analyzed)</u>: *Whole commodity as prepared for wholesale or retail distribution.*

Group 066   Teas

<u>Code No.</u>   <u>Commodity</u>

DT 171   **Teas (Tea and Herb teas)**

DT 1110   **Camomile or Chamomile**
              - *Matricaria recutita* L.;
                syn: M. *chamomilla* auct.
              - *Chamaemelum nobile* (L.) All ;
                syn: *Anthemis nobilis* L.

DT 5277   **Camomile, German or Scented,** see Camomile
              Matricaria recutita L.;
              syn: M. *chamomilla* auct.

DT 5279   **Camomile, Roman or Noble,** see Camomile
              Chamaemelum nobile (L.) All.;
              syn: *Anthemis nobilis* L.

DT 1111   **Lemon verbena (dry leaves)**
              Lippia citriodora H.B. & K.;
              syn: L. *triphylla* L'Herb.

DT 1112   **Lime blossoms**
            Tilia cordata Mill., syn: T *ulmifolia* Scop.;
            T *parvifolia* Ehrh. ex Hoffm., *Tilia platyphyllos* Scop.;
            syn: T. *grandifolia* Ehrh. ex Hoffm.

DT 1113   **Maté (dry leaves)**
            Ilex paraguensis D. Don.;
            syn: I. *paraguariensis* St. Hill.

DT 5281   **Mayweed, Scented,** see Camomile, German

DT 5283   **Paraguay tea,** see Maté

DT 5285   **Peppermint tea** (succulent or dry leaves), see Peppermint, Group 027: Herbs

DT 446   **Roselle** (calyx and flowers), **dry**
            Hibiscus sabdariffa L.

DT 1114   **Tea, Green, Black** (black, fermented and dried)
            Camellia sinensis (L.) O Kuntze, several cultivars;
            syn: C. *thea* Link; C *theifera* Griff.; *Thea sinensis* L.;
            T. *bohea*, L. T. *viridis* L.

Vegetable oils, crude

Class D
**Type 13   Derived edible products of plant origin
                  Group 067 Group Letter Code OC**

Group 067. Vegetable oils, crude, includes the crude vegetable oils derived from oil seed Group 023, tropical and sub-tropical oil-containing fruits such as olives, and some pulses (e.g. soya bean, dry). For the definition and characteristics of Olive oil, crude see Codex Stan. 33-1981. The crude oils are used as constituents of compounded animal feeds or further processed (refined, clarified). See Group 068, Vegetable oils, edible (or refined).

Exposure to pesticides is through pre-harvest treatment of the relevant crops or post-harvest treatment of the oilseeds or oil-containing pulses.

Portion of commodity to which the MRL applies (and which is analyzed): *Whole commodity as prepared for wholesale distribution.*

Group 067   Vegetable oils, crude

Code No.   Commodity

OC 172   **Vegetable oils, crude**

OC 5289   **Corn oil, crude,** see Maize oil, crude

OC 691   **Cotton seed oil, crude**

OC 665   **Coconut oil, crude**

OC 645   **Maize oil, crude**

OC 305   **Olive oil, virgin,** see definition in Codex Stan. 33-1981

OC 696   **Palm oil, crude**
                  made from the fleshy fruit mesocarp of *Elaeis guineensis*
                  Jacq., see Codex Stan. 125-1981.

OC 1240   **Palm kernel oil, crude**
                  made from the kernels of the fruits of *Elais guineensis*
                  Jacq., see Codex Stan. 126-1981.

OC 697   **Peanut oil, crude**

OC 495   **Rape seed oil, crude**

OC 699   **Safflower seed oil, crude**

OC 700   **Sesame seed oil, crude**

OC 541   **Soya bean oil, crude**

OC 702   **Sunflower seed oil, crude**

Vegetable oils, edible (or refined)

Class D
Type 13   Derived edible products of plant origin
              Group 068 Group Letter Code 0R

Group 068. Vegetable oils, edible (or refined) includes the vegetable oil derived from oil seed, Group 023, tropical and sub-tropical oil-containing fruits such as olives, and some pulses with a high oil content. The edible oils are derived from the crude oils though a refining and/or clarifying process. For definitions and characteristics of the edible oils listed below, see Codex Stan. 20-27 (inclusive), 33, 124 and 126 (inclusive) - 1981.

Exposure to pesticides is through pre-harvest treatment of the relevant crops, or post-harvest treatment of the oilseeds and oil containing pulses.

<u>Portion of commodity to which the MRL applies (and which is analyzed)</u>: *Whole commodity as prepared for wholesale or retail distribution.*

Group 068   Vegetable oils, edible (or refined)

<u>Code No.</u>	<u>Commodity</u>
OR 172	**Vegetable oils, edible**
OR 5291	**Corn oil, edible,** see Maize oil, edible
OR 691	**Cotton seed oil, edible**
OR 665	**Coconut oil, refined**
OR 645	**Maize oil, edible**
OR 305	**Olive oil, refined,** as defined in Codex Stan. 33-1981
OR 5330	**Olive, residue oil,** as defined in Codex Stan. 33-1981, see Olive oil, refined
OR 696	**Palm oil, edible**
OR 1240	**Palm kernel oil, edible**
OR 697	**Peanut oil, edible**
OR 495	**Rape seed oil, edible**

OR 699   **Safflower seed oil, edible**

OR 700   **Sesame seed oil, edible**

OR 541   **Soya bean oil, refined**

OR 702   **Sunflower seed oil, edible**

Miscellaneous derived edible products of plant origin

<u>Class D</u>

Type 13   Derived edible products of plant origin

Group 069   Group Letter Code DM

Group 069. Miscellaneous derived edible products include various intermediate products in the manufacture of edible food products. Some of these are used for further processing and not consumed as food or feed as such.

<u>Portion of the commodity to which the MRL applies (and which is analyzed)</u>: *Whole commodity.*

<u>Note</u>: In view of the number of related commodities in this group extra code numbers had to be used, not related to the primary food commodity from which the intermediate product is prepared. Reference to the additional code numbers is given with the primary food commodity where it is listed for the first time in the Classification.

Group 069   Miscellaneous derived edible products of plant origin

<u>Code No.</u>	<u>Commodity</u>
DM 1	**Citrus molasses**
DM 1215	**Cocoa butter**
DM 1216	**Cocoa mass**
DM 715	**Cocoa powder**
DM 305	**Olives, processed**
DM 658	**Sorghum molasses**
DM 596	**Sugar beet molasses**
DM 659	**Sugar cane molasses**

Fruit juices

<u>Class D</u>
**Type 13   Derived edible products of plant origin
            Group 070 Group Letter Code JF**

Fruit juices, Group 070, are pressed from various mature fruits, either from the whole fruits or from the pulp (Type 1 and fruits from fruiting vegetables, Groups 011 and 012). A small amount of preserving agent(s) may be added to the juices during processing. The juices are often prepared for international trade in a concentrated form which is reconstituted for wholesale or retail distribution to about the original juice concentration as obtained by the pressing process.

<u>Portion of the commodity to which the MRL applies (and which is analyzed)</u>: *Whole commodity (not concentrated) or commodity reconstituted to the original juice concentration.*

Group 070   Fruit juices

<u>Code No.</u>	<u>Commodity</u>
JF 175	**Fruit juices**
JF 1	**Citrus juice**
JF 4	**Orange juice**
JF 226	**Apple juice**
JF 5293	**Cassis, see Black currant juice**
JF 1140	**Black currant juice**
JF 269	**Grape juice**
JF 203	**Grapefruit juice**
JF 341	**Pineapple juice**
JF 448	**Tomato juice**

By-products, used for animal feeding purposes, derived from fruit and vegetable processing

<u>Class D</u>
Type 13   Derived edible products of plant origin
Group 071   Group Letter Code AB

Group 071. The commodities of this group are by-products derived from fruit and vegetable processing which are mainly used for animal feeding purposes either as a part of the ration of livestock animals as such, or as an element in the manufacture of compounded feeds. The commodities are prepared, in general, in a dry form for wholesale or retail distribution.

<u>Portion of the commodity to which the MRL applies (and which is analyzed)</u>: *Whole commodity. Residues in "wet" commodities of this group should be expressed on a "dry-weight" basis; see explanation in Group 050, Legume animal feeds.*

Group 071   By-products, used for animal feeding purposes, derived from fruit and vegetable processing

<u>Code No.</u>	<u>Commodity</u>
AB 1	**Citrus pulp, dry** *Citrus* spp.
AB 226	**Apple pomace, dry** *Malus domesticus* Borkhausen
AB 269	**Grape pomace, dry** *Vitis vinifera* L.
AB 596	**Sugar beet pulp, dry** *Beta vulgaris* L., var. *saccharifera*; syn: B. *vulgaris* L., var. *altissima*
AB 1201	**Sugar beet pulp, wet** Residues in the wet pulp to be expressed on a dry weight basis.

Manufactured Foods (single-ingredient) of plant origin

<u>Class D</u>
Type 14   Manufactured Foods (single-ingredient) of plant origin

The term "single-ingredient manufactured food" means a "processed food" which consists of one identifiable food ingredient, with or without packing medium or minor ingredients, such as flavouring agents, spices and condiments, and which is normally pre-packaged and ready for consumption with or without cooking.

Manufactured foods (multi-ingredient) of plant origin

<u>Class D</u>
Type 15   Manufactured foods (multi-ingredient) of plant origin

The term "multi-ingredient manufactured food" means a processed food, consisting of more than one major ingredient.

A multi-ingredient food consisting of ingredients of both plant and animal origin will be included in this type if the ingredient(s) of plant origin is (are) predominant.

Manufactured multi-ingredient cereal products

<u>Class D</u>
Type 15   Manufactured foods (multi-ingredient) of plant origin
               Group 078 Group Letter Code CP

Group 078. Manufactured multi-ingredient cereal products.

The commodities of this group are manufactured with several ingredients; products derived from cereal grains however form the major ingredient.

<u>Portion of the commodity to which the MRL applies (and which is analyzed)</u>: *Whole commodity as prepared for wholesale or retail distribution.*

Group 078   **Manufactured multi-ingredient cereal products**

<u>**Code No.**</u>   <u>**Commodity**</u>

CP 179   **Bread and other cooked cereal products**

CP 5295   **Corn bread,** see Maize bread

CP 645   **Maize bread**

CP 1250   **Rye bread**

CP 1211   **White bread**

CP 1212   **Wholemeal bread**

CLASS E   PROCESSED FOODS OF ANIMAL ORIGIN

Definition, see Class D.

Class E

Type 16   Secondary food commodities of animal origin

The term "secondary food commodity" means a "primary food commodity" which has undergone simple processing, such as removal of certain portions, drying, and comminution, which do not basically alter the composition or identity of the commodity.

Secondary food commodities may be processed further, or used as ingredients in the manufacture of food, or sold directly to the consumer.

This type of processed food includes groups of processed primary food commodities of animal origin which have undergone simple processing, such as processed mammalian meat and poultry meat, fishes and other aquatic animals, e.g. dried meat, dried fish.

Dried meat and fish products

Class E
**Type 16   Secondary food commodities of animal origin
Group 080 Group Letter Code MD**

Group 080. Dried meat and fish products, includes natural or artificial dried meat products and dried fishes, mainly marine fishes. Most of the dried fishes are naturally dried (wind and sun). For convenience other marine animals, whether or not fishes or Crustaceans, are classified in this group.

The entire commodity may be consumed, either as such or after processing (c.q. dried fish).

Portion of commodity to which the MRL applies (and which is analyzed): *Whole commodity as prepared for wholesale or retail distribution.*

Group 080   Dried meat and fish products

Code No.	Commodity
MD 95	**Meat, dried** (from mammals other than marine mammals)
MD 180	**Dried fish**
MD 120	**Diadromous fish, dried**

MD 127   **Flat-fishes, dried**
                (See Group 042, Subgroup 0127)

MD 125   **Marine fish, dried**

MD 126   **Stockfish (= dried Cod and Cod-like fishes)**
                (see Group 042, Subgroup WS 0126)

MD 812   **Cattle meat, dried** (including dried and smoked)

MD 816   **Horse meat, dried** (including dried and smoked)

MD 818   **Pig meat, dried** (including dried and smoked)

MD 5297   **Beche-de-mer, dried,** see Sea-cucumbers, dried

MD 927   **Cod, dried**

MD 929   **Dab or Common dab, dried**

MD 935   **Hakes, dried**

MD 936   **Halibut, dried**

MD 940   **Ling, dried**

MD 1010   **Sea-cucumbers, dried**

Class E
Type 17   **Derived edible products of animal origin**

The term "Derived edible products" means foods or edible substances isolated from primary food commodities or raw agricultural commodities not intended for human consumption as such, using physical, biological and chemical processes.

This type includes processed (rendered or extracted, possibly refined and/or clarified) fats from mammals, including aquatic mammals, poultry and aquatic organisms such as fishes.

Secondary milk products

<u>Class E</u>

**Type 16   Secondary Food Commodities of animal origin
                Group 082 Group Letter Code LS**

Group 082, secondary milk products, includes milk products which have undergone simple processing such as removal or part removal of certain ingredients e.g. water, milk fat etc. The group and the commodities therein will only be used for pesticides which are not partitioned exclusively or nearly exclusively into the milk fat.

The recommended system for expressing the MRLs for the fat-soluble pesticides in milk and milk products is explained in the introduction to CAC/VOL. XIII-Ed.2. The group includes among others the following commodities as defined in the relevant Codex Standards, see CAC/VOL. XVI-Ed.1 (1984). The reference no. of the standard is indicated between brackets.

Milk powders (whole, skimmed and partly skimmed) (Standard A-5 1971); evaporated milks (whole, skimmed) (Standard A-3 1971); skimmed milk.

Specific commodities will be listed in this group with their code nos. accordingly as the necessity for this arises.

<u>Portion of the commodity to which the MRL applies (and which is analyzed)</u>: *Whole commodity*.

Crustaceans, processed

<u>Class E</u>

**Type 17   Derived edible products of animal origin
                Group 084 Group Letter Code SC**

Group 084, Crustaceans, processed. Crustaceans are processed to a large extent before entering the national or international trade channels.

Crabs, lobsters and shrimps or prawns are in general cooked directly after catching. Thereafter either the animals are deep-frozen with or without shell, or the meat without shell is canned, with or without a packing medium. The latter may consist of water, salt, lemon juice and sugars.

Shrimps or prawns may also be "parboiled" and thereafter deep-frozen.

According to the relevant Codex Standards, namely 92-1981, and 95-1981 "cooked" means heated for a period of time such that the thermal centre reaches a temperature adequate to coagulate the protein and "parboiled" means heated for a period of such time that the surface of the product reaches a temperature adequate to coagulate the protein at the surface but inadequate to coagulate the protein at the thermal centre.

The cooked commodities are in general subjected to deep-freezing directly after cooking or the cooking is part of the canning process.

The designation cooked after the commodity may include any of the processes mentioned except the parboiled and deep-frozen shrimps or prawns.

The entire commodity except the shell may be consumed.

Portion of the commodity to which the MRL applies (and which is analyzed): *Whole commodity (especially with the small sized species) or the cooked meat without shell as prepared for wholesale or retail distribution.*

For commodity description and scientific family or species names see Group 045, Crustaceans.

Group 084   Crustaceans, processed

Code No.	Commodity
SC 143	**Crustaceans, cooked**
SC 144	**Freshwater crustaceans, cooked**
SC 145	**Marine crustaceans, cooked**
SC 146	**Crabmeat, cooked**
SC 976	**Freshwater crayfishes, cooked**
SC 977	**Freshwater shrimps or prawns, cooked**
SC 978	**Lobsters (including Lobster meat), cooked**
SC 979	**Shrimps or Prawns, cooked**
SC 1220	**Shrimps or Prawns, parboiled**

Animal fats, processed

<u>Class E</u>
Type 17   Derived edible products of animal origin
**   Group 085 Group Letter Code FA**

The Group 085, processed animal fats, includes rendered or extracted (possibly refined and/or clarified) fats from land and aquatic mammals and poultry and fats and oils derived from fishes.

<u>Portion of the commodity to which the MRL applies (and which is analyzed)</u>: *Whole commodity as prepared for wholesale or retail distribution.*

Group 085   Animals fats, processed

<u>Code No.</u>   <u>Commodity</u>

FA   96   **Tallow and lard from cattle, goats, horses, pigs and sheep**
                    (lard only from pigs)

FA   111   **Poultry fats, processed**

FA   142   **Processed Fat (Blubber), of Whales, Dolphins and Seals**

FA   810   **Buffalo tallow**
                    Bubalis bubalis L.;
                    Syncerus caffer Sparrman
                    Bison bison L.

FA   811   **Camel tallow**
                    Camelus bactrianus L.; C. *dromedarius* L.;
                    Lama glama L.; L *pacos* L.

FA   812   **Cattle tallow** (including processed suet)
                    Bos taurus L.;
                    see further Group 030, no. MM 0812

FA   814   **Goat tallow**
                    Capra hircus L.; other *Capra* spp.

FA   816   **Horse tallow**
                    Equus caballus L.

FA 818   **Lard** (of pigs)
            among others *Sus domesticus* Erxleben; other *Sus* spp. and ssp.

FA 822   **Sheep tallow**
            Ovis aries L.; other *Ovis* spp.

FA 972   **Whales, Blubber of, processed**

FA 840   **Chicken fat, processed**
            Gallus gallus L.; other *Gallus* spp.

FA 841   **Duck fat, processed**
            Anas platyrhynchos L.; other *Anas* spp.

FA 842   **Goose Fat, processed**
            Anser anser L.; other *Anser* spp.

FA 848   **Turkey Fat, processed**
            Meleagris gallopavo L.

Milk fats

<u>Class E</u>

Type 17   Derived edible products of animal origin
**   Group 086 Group Letter Code FM**

Group 086. Milk fats are the fatty ingredients derived from the milk of various mammals.

<u>Portion of the commodity to which the MRL applies and which is analyzed)</u>: ***Whole commodity.***

Group 086   Milk fats

<u>Code No.</u>   <u>Commodity</u>

FM 183   **Milk fats**
            (from milk of Buffalo, Camel, Cattle, Goat or Sheep)

FM 810   **Buffalo milk fat**
            Bubalis, bubalis L.;
            Syncerus caffer Sparrman: *Bison bison* L.

FM 811   **Camel milk fat**
            Camelus bactrianus L. C. *dromedarius* L,;
            Lama glama L.; L. *pacos* L.

FM 812   **Cattle milk fat**
            Bos taurus L.; see further Group 030 no. ML 0812

FM 814   **Goat milk fat**
            Capra hircus L., other Capra spp.

FM 822   **Sheep milk fat**
            Ovis aries L.; other *Ovis* spp.

Group 087   Derived Milk products

<u>Class E</u>
Type 17   Derived edible products of animal origin
                  Group 087 Group Letter Code LD

Group 087 includes food or edible substances isolated from the primary food commodity cattle milk, or milks from other mammals, using physical, biological and chemical processes. This group and the commodities therein will only be used if necessary for pesticides which are not partitioned exclusively or nearly exclusively into the milk fat. For further explanation, see Group 082.

This group includes among others the following food commodities, as defined in the relevant Codex Standards, see CAC/VOL. XXI, Ed-1 (1984): Butter, whey butter, both in Standard A-1 (1971), Butteroil, anhydrous butteroil, both in Standard A-2 (1973); Cream, Standard A-9 (1976); Cream powders (half cream, high fat), Standard A-10 (1971); Edible acid casein, Standard A-12 (1976): Edible caseinates, Standard A-13 (1976).

Specific commodities will be listed in this group with their code nos. accordingly as the necessity for this arises.

<u>Portion of the commodity to which the MRL applies (and which is analyzed)</u>: *Whole commodity*.

Manufactured food (single-ingredient) of animal origin

<u>Class E</u>

Type 18   Manufactured food (single ingredient) of animal origin

The term "single ingredient manufactured food" means a processed food which consists of one identifiable food ingredient, with or without packing medium or minor ingredients such as flavouring agents, spices and condiments, and which is normally pre-packaged and ready for consumption, with or without cooking.

Group 090   Manufactured milk products (single ingredient)

<u>Class E</u>

Type 18   Manufactured food (single ingredient) of animal origin
**   Group 90 Group Letter Code LI**

Group 090 and the commodities therein will only be used, if the necessity arises, for pesticides which are not partitioned exclusively or nearly exclusively into the milk fat. For further explanation see Group 082.

This group includes among others the following food commodities, as defined in the relevant Codex Standards (indicated between brackets); Yoghurt (Codex Standard A-11(a) 1975); Cheeses, individually named (Codex Standard A-6 1978 and Standard C-1 (1966-1978).

Specific commodities will be listed in this group with their code nos. accordingly as the necessity for this arises.

<u>Portion of the commodity to which the MRL applies (and which is analyzed)</u>: *Whole commodity as prepared for wholesale or retail distribution.*

Manufactured food (multi-ingredient) of animal origin

<u>Class E</u>
Type 19   Manufactured food (multi-ingredient) of animal origin

The term "multi-ingredient manufactured food" means a processed food consisting of more than one major ingredient.

A multi-ingredient food consisting of ingredients of both animal and plant origin will be included in this type if the ingredient(s) of animal origin is (are) predominant.

Group 092   Manufactured milk products (multi-ingredient)

<u>Class E</u>
Type 19   Manufactured food (multi-ingredient) of animal origin
         Group 092   Group Letter Code LM

Group 092 and the commodities therein will only be used in the classification if necessary for pesticides which are not partitioned exclusively or nearly exclusively into the milk fat. For further explanation see Group 082.

This group includes among others the following commodities, as defined in the relevant Codex Standards, see CAC/VOL.XVI, Ed-1 (1984); Processed Cheese Products, Codex Standard A-8(a) and A-8(b) (1978) Processed Cheese Preparations, Standard A-8(c) (1978); Flavoured Yoghurt, Standard A-11(b) (1976); Sweetened Condensed Milk, Standard A-4 (1971).

Specific commodities will be listed in this group with their code nos. accordingly as the necessity for this arises.

<u>Portion of the commodity to which the MRL applies (and which is analyzed)</u>: *Whole commodity as prepared for wholesale or retail distribution.*

BIBLIOGRAPHY

In addition to several taxonomic handbooks and guides dealing with specific botanical and zoological families, the following references among others were consulted.

Australia, Dept. of Primary Industry, 1981
    Definitions and Classification of Food and Food Groups,
    Document PB 413
    Australian Government Publishing Service, Canberra

Bailey, L.A., 1958, The Standard Cyclopedia of Horticulture,
    2nd edition, 17th printing 3 vols.
    The MacMillan Co., New York

CIBA-GEIGY, 1975 CITRUS, Häfliger E., Ed. Technical Monograph No. 4.
    CIBA-GEIGY Agrochemicals, Basle Switzerland 1-88

Codex Alimentarius Commission
    C.A.C. 1981a Codex Standards for Processed Fruits and Vegetables
    and Edible Fungi
    CAC/Vol. II - Ed. 1
    Joint FAO/WH0 Food Standards Programme, Rome

Codex Alimentarius Commission
    C.A.C. 1981b Codex Standards for Processed Meat and Poultry Products and Soups
    and Broths
    CAC/Vol. IV - Ed. 1
    Joint FAO/WHO Food Standards Programme, Rome

Codex Alimentarius Commission
    C.A.C. 1981c Codex Standards for Fish and Fishery Products,
    CAC/Vol. V - Ed. 1
    Joint FAO/WHO Food Standards Programme, Rome

Codex Alimentarius Commission
    C.A.C. 1981d Codex Standards For Cocoa Products and Chocolate
    CAC/Vol. VII - Ed. 1
    Joint FAO/WHO Food Standards Programme, Rome

Codex Alimentarius Commission
    C.A.C. 1981e Codex Standards for Quick-frozen Fruits and Vegetables
    CAC/Vol. VIII - Ed. 1
    Joint FAO/WHO Food Standards Programme, Rome

Codex Alimentarius Commission
        C.A.C. 1984 Code of Principles Concerning Milk and Milk Products,
        8th edition. International Standards for Milk Products and International Individual
        Standards for Cheeses
        CAC/Vol. XVI, Ed - 1.

Codex Alimentarius Commission
        C.A.C./RCP 7-1974 (1975) Recommended International System for the Description
        of Carcases of Bovine and Porcine Species and Recommended International
        Description of Cutting Methods of Commercial Units of Beef, Veal, Lamb and
        Mutton, and Pork Moving in International Trade
        Joint FAO/WHO Food Standards Programme, Rome

FAO 1981, Yearbook of Fishery Statistics Vol. 50 and 58
        1984 FAO Fisheries Series No. 16 and 25
        FAO Statistics Series No. 38 and 65, FAO, Rome

FAO Agris 1979, Codebook Agris Terminology and Codes
        FAO, Rome (see also Prince-Perciballi 1983)

FAO 1983 Plants and Plant Products FAO Terminology Bulletin 25/1 pp. 1-328

Göhl, R., 1981, Tropical Feeds
        FAO Animal Production and Health Series No. 12
        FAO, Rome

Magness, J.R., Markle, G.M., and Compton, C.C., 1971
        Food and Feed Crops of the United States I.R. Bulletin No. 1
        New Jersey Agr. Exp. Sta. College of Agriculture and Environmental Sciences,
        Rutgers University, The State University of New Jersey, New Brunswick

Mason, I.L., 1984, Evolution of Domesticated Animals
        Longman, London, New York

Prince-Perciballi, I., 1983, Agris/Caris Categorization
        Scheme FAO/Agris - 3 (Rev. 4)

Pursglove. J.W., 1957-1988
        Tropical Crops Dicotyledons 1-719
        Tropical Crops Monocotyledons 1-607
        Longman, London U.K.

Reuther, W., Webber, H.J., and Batchelor, L.D.. Eds. 1967
        The Citrus Industry Vol. 1 Revised Edition
        University of California, Division of Agricultural Sciences

Sánchez - Monge y Parellade, E., 1980
    Dicionario de Plantas Agrícolas
    Ministerio de Agricultura, Servicio de Publicaciones Agrarias
    pp. 1-328

Tanaka, T., 1976, Tanaka's Cyclopedia of Edible Plants of the World: Nakao S., Ed.
    Keigako Publishing Co., Tokyo Japan

Terrel, E.E. 1977
    A checklist of names for 3000 vascular plants of economic importance. Agricultural
    Handbook No. 505 Agricultural Research Service USDA, Washington, D.C.

Tidbury, C.E., 1983, CAB Thesaurus Vol. 1 (A-1)
    Commonwealth Agricultural Bureaux, Slough, England

USA 1983, Code of Federal Regulations
    Title 40 Protection of the Environment Part 18, par. 180, 134.
    Tests on the amounts of residues remaining
    Federal Register, Special Edition

Westphal, E., 1982, Tropical Pulses
    Agricultural University Wageningen, The Netherlands.
    Laboratory of Tropical Plant-breeding

Zeven, A.C., and de Wet, J.M.J., 1982
    Dictionary of Cultivated Plants and Their Region of Diversity
    Pudoc, Centre for Agricultural Publishing and Documentation, Wageningen, The
    Netherlands

SECTION 3

RECOMMENDED METHODS OF SAMPLING
FOR THE
DETERMINATION OF PESTICIDE RESIDUES

RECOMMENDED METHODS OF SAMPLING
FOR THE
DETERMINATION OF PESTICIDE RESIDUES

INTRODUCTION

This Section contains two recommended methods of sampling food products and animal feeds, for the determination of their pesticide residue content. The first recommended method covers all commodities except meat and poultry products. The second recommended method is devoted to those products only. The reason for this division in sampling procedures relates to the way Codex MRLs for these products are established, and is discussed below under the heading, "Basis for the Sampling Procedures".

BASIS FOR THE SAMPLING PRINCIPLE

Sampling for the enforcement of an MRL should be consistent with principles applied in the setting of that MRL, and must be practical for the examination of lots in trade.

MRLs for meat and poultry products are developed from experimental residue data obtained in field trials where animals are treated or exposed to the pesticide in accordance with good agricultural practice (GAP). In these experiments various edible tissues from individual livestock and poultry are separately analyzed, except when the combining of tissue from more than one animal is required to obtain an adequate sample size for analysis (e.g., for poultry organs). The Joint Meeting for Pesticide Residues (JMPR) evaluates the residue data and recommends an MRL consistent with national GAPs that is not expected to be exceeded in any animal when marketed for human food.

For most other commodities, including eggs and milk, the recommended sampling for field trials involves collection of a bulk sample made up of a number of primary samples which are combined as the final sample. The final sample, or a representative part, is then analyzed. The JMPR evaluates the residue data from these final samples and recommends an MRL consistent with GAP that is not expected to be exceeded in the raw agricultural commodity when marketed.

Thus, the principle of applying an MRL for meat and poultry products to the residue concentration found in primary samples, and applying the MRL for most other commodities to the residue concentration found in a "final sample" is consistent with the data evaluation used by the JMPR in recommending MRLs to the Codex Committee on Pesticide Residues.

COMPATIBILITY WITH NATIONAL RESIDUE CONTROL PROGRAMMES

It is important to emphasize that for effective residue control in meat and poultry products intended for export, sampling should occur at the time of slaughter before the product is packaged or further processed for commerce. Only at slaughter are fresh target tissues routinely available for determining the presence of residues. There is also a greater likelihood of sampling animals which have been raised under uniform conditions, and thus with more uniform exposure to a pesticide. This allows findings to be extrapolated to the larger population. Sampling at point of entry of packaged meat products should be designed for quality assurance purposes in monitoring the effectiveness of a member country's domestic residue control programme, but should not be viewed as the most effective means of controlling pesticide residues.

While the interest of Codex is in the examination of products in international trade (i.e., sampling for enforcement purposes by an importing country) it is desirable for Codex recommendations to be consistent in principle and appropriate for use by countries in their domestic control programmes as well. Such consistency as regards sampling avoids the dilemma some countries may face when national legislation requires that the same standards be applied to domestically-produced and imported products.

Many countries sample animals at slaughter for residue testing, and when violative residues are found, they use animal source traceback, quarantine, or other methods to prevent marketing of additional animals until testing indicates the identified problem has been corrected. These very effective control programs are based on testing of primary samples. By adopting the principle of applying the Codex MRL to a primary sample, uniformity can be achieved in the application of MRLs by exporting countries that carry out such testing programmes and by the importing country. This uniformity is particularly important for countries that accept imported meat products based in part on the evaluation of the effectiveness of the residue control and testing programmes conducted by the exporting country.

APPLICATION OF THE SAMPLING PRINCIPLE

In both Recommended Methods of Sampling a lot is defined as, "an identifiable quantity of goods delivered at one time, having or presumed by the sampling officer to have common properties or uniform characteristics such as the same origin, the same variety, the same consigner, the same packer, the same type of packing or the same mark." The sampling officer must determine from information at hand what quantity of material represents a lot. In the absence of producer codes, a consignment frequently is treated as a lot, even though it comprises products from different locations produced under non-uniform conditions of exposure to pesticides.

Under the recommended sampling principles, a lot would comply with the MRL if (a) the final sample (consisting of combined primary samples) of commodities other than meat and poultry products did not contain a residue above the MRL, or (b) none of the primary samples of meat and poultry products analyzed contained a residue above the MRL. If some, but not all, of the primary samples complied with the MRL, these results would indicate that some units in the "lot" had been exposed to the pesticide under conditions that did not comply with GAP. Such a "lot" would represent commingling of contaminated and noncontaminated products. While it may be possible by sublotting and additional testing to separate out the parts that complied with the MRL, an importing country should not be required to assume this burden.

Sampling design

A different approach and level of sampling is recommended to be used for lots when there is reason to believe that food may not be in compliance with the MRL (i.e., "suspect" lots) from that to be used for lots when there is no reason to believe the food may not be in compliance with the MRL (i.e., "non-suspect" lots). A lot may be "suspect", for example, because it came from a source with a history of non-compliance with MRLs, where there is evidence that contamination during transport may have occurred, when inspection of live animals imported for slaughter reveals signs of toxicosis, or when other relevant information is available to the inspection official.

Sampling of non-suspect lots

A statistically-based random sampling programme is recommended for non-suspect lots, which typically draws primary samples from many lots throughout the year with a minimum of sampling from any one lot. Examples are discussed below:

a. Stratified random sampling

Samples are obtained by separating the population elements into some non-overlapping groups, called strata, and selecting samples within each stratum according to a simple random design. Countries, or geographic regions, are natural strata because agricultural practices are likely to be more uniform, tending to make products within these groups more alike. It is also common to stratify by time (e.g., month, quarter) for convenience and efficient use of resources, and to detect seasonal variations. Tables of random numbers or equivalent procedures are used to ensure randomization. However, even with a computer network the simple random design criteria are mechanistically difficult to apply when commodities must be sampled at many different locations over an extended time period.

b. Systematic sampling

An example of systematic sampling is taking a sample from every "X" kilos of product imported from a particular country. This method is convenient when there is reliable information on product volumes that can be used to determine the sampling interval that will give the desired number of samples per month or year. Alternatively, samples may be systematically taken by time or number of shipments. As systematic sampling can be vulnerable to abuse if the system is predictable, it is advisable to build some randomness into the sampling interval.

c. Biased, or estimated worst-case sampling

This design is useful when a population group anticipated to be at greatest risk can be identified. For example, a production class of animals or other food products from certain regions may be randomly sampled during a particular season when agricultrual practices favour use of certain pesticide chemicals.

These designs provide a method for testing imported products to identify types of products and sources that do not comply with Codex MRLs and, therefore, may warrant more intensive examination of future shipments or regulatory follow-up if the sampled lot can be located. Some of the designs may allow estimation of the extent to which imported products as a whole comply with Codex MRLs. Table 1 below provides statistical information relevant to deciding the number of samples to select, which national authorities may consider in relation to source constraints for systematic testing of compliance with Codex MRLs.

TABLE 1

Number of samples required to detect at least one violation with predefined probabilities (i.e., 90, 95 and 99 percent) in a population having a known violation incidence rate.

Violation Incidence (%) in a Population	Minimum number of samples (n_o) required to detect a violation with a confidence of:		
	90%	95%	99%
35	6	7	11
30	7	9	13
25	9	11	17
20	11	14	21
15	15	19	29
10	22	29	44
5	45	59	90
1	230	299	459
.5	460	598	919
.1	2302	2995	4603

The number of primary samples does not depend on population size, except when the number of samples shown in the table is greater than about 10% of the population size. The following formula can be used to adjust the table values for the minimum number of primary samples (n_o) and compute the required minimum number of primary samples (n) for a given lot size (N):[1]

$$n = \frac{n_o}{1 + (n_o - 1)/N}$$

Sampling of suspect lots

It is recommended that at least 6 and usually no more than 30 primary samples be analyzed from a suspect lot. The smaller number of samples would be appropriate, for example, when the suspected contamination is likely to occur throughout the lot, or when the location of probable contamination (e.g., surface contamination) is readily identified. Table 1 provides statistical information that may be helpful in deciding the number of samples to be analyzed in a particular case. International harmony in control procedures is not dependent on the number of primary samples analyzed because the MRL is applied to each primary sample. However, as shown in Table 1, the larger the number of samples taken, the greater the assurance that products not in compliance will be detected.

---

[1] *Cochran, William G., Sampling Techniques, 2nd Ed., 1963, pp. 74-75, John Wiley and Sons, Inc.*

RECOMMENDED METHOD OF SAMPLING PRODUCTS
OTHER THAN MEAT AND POULTRY
FOR THE DETERMINATION OF PESTICIDE RESIDUES

1. OBJECTIVES

For the examination of a lot in order to discover whether it complies with Codex Maximum Limits for Pesticide Residues it is necessary to provide a representative sample for analysis. The objective of this sampling procedure is to obtain a _final sample_ representative of the _lot_ in order to determine its average pesticide residue content. The _final sample_ is considered representative of the _lot_ in order to determine its average pesticide residue content. The _final sample_ is considered representative of the _lot_ when the procedure outlined below has been followed. The Codex Maximum Residue Limit applies to the _final sample_.

2. DEFINITIONS

2.1 Lot [1]

A identifiable quantity of goods delivered at one time, having or presumed by the sampling officer to have common properties or uniform characteristics such as the same origin, the same variety, the same consigner, the same packer, the same type of packing or the same mark. Several lots may make up a consignment.

2.2 Consignment

A quantity of material covered by a particular consignment note or shipping document. _Lots_ in the same consignment may be delivered at different times and may have different amounts of pesticide residues.

2.3 Primary Sample

A quantity of material taken from a single place in the _lot_.

2.4 Bulk Sample

Combined total of all the _primary samples_ taken from the same _lot_.

2.5 Final Sample

Bulk sample or representative part of the _bulk sample_ to be used for control purposes.

2.6 Laboratory Sample

Sample intended for the laboratory. The final sample may be used as a whole or subdivided into representative portions (laboratory sample) if required by national legislation.

---

[1] _The identification of a lot would be greatly facilitated by the use of farmer and packer codes._

3.   **EMPLOYMENT OF AUTHORIZED SAMPLING OFFICERS**

The samples must be taken by officers authorized for the purpose by the appropriate authorities.

4.   **SAMPLING PROCEDURE**

4.1   Material to be sampled

Each lot which is to be examined must be sampled separately.

4.2   Precautions to be Taken

In the course of taking the primary samples and in all subsequent procedures precautions must be taken to avoid contamination of the samples or any other changes which would adversely affect the amount of residues or the analytical determinations or make the laboratory sample not representative of bulk sample.

4.3   Primary Samples

As far as practicable these should be taken throughout the lot. Departures from this requirement must be recorded (see para. 7). As far as possible the primary samples should be of similar size and the combined total of all the primary samples (bulk sample) must not be less than that required for the final sample bearing in mind the possible requirement of further subdivision and the provision of adequate laboratory samples. The minimum number of primary samples to be taken is given in the following Table:

Weight of lot in kilogrammes	Minimum number of primary samples to be taken
< 50	3
51 - 500	5
501 - 2000	10
> 2000 (*)	15

(*)   For whole cereals and other materials shipped in bulk, well established alternative sampling procedures are available and may be used providing these are recorded (see para 7) and the minimum requirements in para 4.6.4 are met.

For processed products in cans, bottles, packages or other small containers, especially when the sampling officer does not know the weight of the lot, the following sampling plan may be followed:

Number of cans, packages or containers in the lot	Minimum number of primary samples to be taken
1 - 25	1
26 - 100	5
101 - 250	10
> 250	15

For homogeneous lots a sample fully representative of the whole is obtained by withdrawing any single sample.

4.4   Preparation of Bulk Sample

The bulk sample is made by uniting and mixing the primary samples.

4.5   Preparation of Final Sample

4.5.1   The bulk sample should, if possible, constitute the final sample.

4.5.2   If the bulk sample is too large the final sample may be prepared from it by a suitable method of reduction. In this process, however, individual fruits and vegetables must not be cut or divided.

4.6   Preparation of the Laboratory Sample

4.6.1   The final sample should, if possible, be submitted to the laboratory for analysis.

4.6.2   If the final sample is too large to be submitted to the laboratory a representative subsample must be prepared.

4.6.3   National legislative needs may require that the final sample be subdivided into two or more portions for separate analysis. Each portion must be representative of the final sample. The precautions in para 4.2 should be observed.

4.6.4   The minimum amount of material to be submitted to the laboratory, i.e., the size of the laboratory sample, is as follows:

Commodity	Examples	Minimum quantity required
Small or light products, unit weight up to 25g	berries } peas } olives } parsley }	1 kg
Medium sized products, unit weight usually between 25 and 250g	apples } oranges } carrots } potatoes }	1 kg (at least 10 units)
Large sized products, unit weight over 250g	cabbage } melons } cucumbers }	2 kg (at least 5 units)
Dairy products	whole milk } cheese } butter } cream }	0.5 kg
Eggs	—	0.5 kg (10 units if whole)
Fat, fish, and other fish and animal products	—	1 kg
Oils and fats	cottonseed } oil } margarine }	0.5 kg
Cereals and cereal products	—	1 kg

5.   PACKAGING AND TRANSMISSION OF LABORATORY SAMPLES

The laboratory sample must be placed in a clean inert container offering adequate protection from external contamination and protection against damage to the sample in transit. The container must then be sealed in such a manner that unauthorized opening is detectable, and sent to the laboratory as soon as possible taking any necessary precautions against leakage or spoilage, e.g., frozen foods should be kept frozen, perishable samples should be kept cooled or frozen.

6.   RECORDS

Each laboratory sample must be correctly identified and should be accompanied by a note giving the nature and origin of the sample and the date and place of sampling, together with any additional information likely to be of assistance to the analyst.

7.   DEPARTURES FROM RECOMMENDED SAMPLING PROCEDURE

If, for any reason, there has had to be a departure from the recommended procedures, especially paragraph 4, full details of the procedure actually followed must be recorded in the accompanying note (see para. 6).

RECOMMENDED METHOD OF SAMPLING
OF MEAT AND POULTRY PRODUCTS
FOR THE DETERMINATION OF PESTICIDE RESIDUES

1.   **OBJECTIVE**

To provide instructions for sampling a lot of meat and poultry products to determine for control purposes whether it complies with Codex Maximum Residue Limits (MRLs).

2.   **DEFINITIONS**

2.1   Lot[1]

An identifiable quantity of food delivered at one time, having or presumed by the sampling officer to have common characteristics, such as the same origin, the same variety, the same packer or consigner, the same type of packing, or the same mark. Several lots may make a consignment.

2.2   Consignment

A quantity of food covered by a particular contractor shipping document. Lots in the consignment may be delivered at different times and have different origins.

2.3   Primary sample

A quantity of food taken from a single animal or place in the lot. Where a single place does not provide a quantity of material adequate for analysis, samples for more than one animal or location are combined for the primary sample (e.g., poultry organs).

2.4   Laboratory sample

Sample intended for the laboratory. The entire primary sample may be used for analysis, or it may be subdivided into representative portions (laboratory samples) if required by national legislation.

3.   **COMMODITIES TO WHICH THE GUIDELINE APPLIES**

The commodity designations listed in 3.1 and 3.2 are in accordance with the Codex Commodity Classifications given in Section 2 of this Volume and its commodity descriptions, except as further defined.

3.1   Selected Class B: Primary Food Commodities of Animal Origin

Type 06 Mammalian Products

No. 030 Meat (Mammalian)
No. 031 Fat (Mammalian)
No. 032 Edible Offal (Mammalian)

---

[1]   *The identification of a lot would be greatly facilitated by the use of farmer and packer codes.*

Type 07 Poultry Products

No. 036 Poultry Meats
No. 037 Poultry Fats
No. 038 Poultry, Edible Offal

3.2 Selected Class E: Processed Products of Animal Origin made only from Primary Foods Nos. 030, 032, 036 and 038.

Type 16 - Secondary Products

Type 18 - Manufactured (single ingredient) products of container or unit size of at least one kilogramme

Type 19 - Manufactured (multiple ingredient) products of container or unit size of at least one kilogramme.

4.   PRINCIPLE APPLIED

The MRL is applied to the residue concentration found in each <u>primary sample</u> taken from a <u>lot</u> for control purposes. A <u>lot</u> complies with a Codex MRL when none of the <u>primary samples</u> contain a residue greater than the MRL.[2]

5.   EMPLOYMENT OF AUTHORIZED SAMPLING OFFICERS

The sample must be taken by officers authorized for the purpose by appropriate authorities.

6.   SAMPLING PROCEDURES

6.1   <u>Material to be sampled</u>

Each <u>lot</u> which is to be examined must be sampled separately.

6.2   <u>Precautions to be taken</u>

In the course of taking the <u>primary samples</u> and in all subsequent procedures, precautions must be taken to avoid contamination of the samples or any other changes which would alter the residue or compromise the analytical determination.

6.3   <u>Collection of a primary sample</u>

Table 1 following paragraph 8, provides detailed instructions for taking a <u>primary sample</u> of the various commodities. The quantity required for laboratory analysis is method dependent; however, the

---

[2]   *When some but not all primary samples comply with a Codex MRL, these results indicate that some units of the lot were treated or exposed under conditions that do not comply with good agricultural practice. While it may be possible by sublotting and further testing to separate out the portion of the lot not in compliance, this burden need not be assumed by the importing country.*

minimum requirements for the <u>laboratory samples</u> listed in Table 1 should be adequate for most analyses. In addition, the following general instructions are provided.

   a.   Whenever possible, each <u>primary sample</u> should be taken from a single animal or unit within a <u>lot</u>, using random selection techniques.

   b.   When a <u>lot</u> derived from imported live animals is sampled on a slaughter line and product from more than one animal is required for adequate sample size, (e.g., for poultry organs) the multiple samples required for the <u>primary sample</u> should be taken as consecutively as practical after random selection of the starting point.

   c.   Canned or packaged products should not be opened for sample unless the unit size is so large that it is impractical to send the whole product to the laboratory. When opening is necessary, the sample should contain a representative portion of liquids surrounding the meat. The sample must then be frozen as described in paragraph 6.5.

   d.   Frozen products should not be thawed before sampling.

   e.   For large units (e.g., prime cuts) containing bone, only a portion of edible tissue should be taken as the <u>primary sample</u>.

6.4   <u>Number of primary samples to be taken from a lot</u>

   It is recommended that a different approach and level of sampling be used for <u>lots</u> when there is reason to believe the food may not be in compliance with MRLs (i.e., "suspect" <u>lots</u>) from that to be used for <u>lots</u> where there is no reason to believe the food may not be in compliance with MRLs (i.e., "non-suspect" <u>lots</u>). A <u>lot</u> may be "suspect", if it originates from a source with a history of non-compliance with MRLs, when there is evidence that contamination during transport may have occurred, when inspection of live animals imported for slaughter reveals signs of toxicosis, or when other relevant information is available to the inspection official.

   6.41   <u>Sampling of suspect lots</u>

   At least 6 <u>primary samples</u> and usually no more than 30 <u>primary samples</u> should be taken. The smaller number of samples is appropriate, for example, when the suspected contamination is likely to occur throughout the <u>lot</u>, or when the location of the probable contamination is readily identified.

   6.42   <u>Sampling of non-suspect lots</u>

   (See 'Application of the Sampling Procedure' and the Table in the Introduction to this Section).

   Some exporting countries conduct comprehensive residue testing programmes and routinely provide results to the importing country. An importing country, therefore, may exempt such products from further testing requirements, or may reduce the level of testing from that normally applied to non-suspect products from other countries that do not provide residue testing results demonstrating compliance with MRLs.

6.5   Packaging and transmission of primary samples

a.   Each primary sample must be placed in a clean inert container offering adequate protection from external contamination and protection against damage to the sample in transit.

b.   The container must then be sealed in such a manner that unauthorized opening is detectable.

c.   The container must be sent to the laboratory as soon as possible after taking precautions against leakage or spoilage.

d.   All perishable samples must be frozen, preferably to minus 20°C, as soon as possible after sample collection. Perishable samples must be transported frozen in a suitable container that retards thawing. If facilities are available, the open container to be used for transporting the samples to the laboratory should be placed in a freezer for 24 hours before packing the pre-frozen sample.[3]

7.   RECORDS

Each primary sample must be correctly identified and should be accompanied by a record giving the nature and country/state/town of origin of the sample, the location at which the sample was taken, the date of sampling, and any additional information likely to be of assistance to the analyst, or to regulatory officials should follow-up action become necessary.

8.   DEPARTURE FROM RECOMMENDED SAMPLING PROCEDURE

If for any reason there has been a departure from the recommended procedures, full details of the procedure actually followed must be recorded in the accompanying records.

---

[3]   *A sample may be placed in a suitable plastic bag. After expelling excess air and closing the top securely, the bagged sample can be placed in a second bag along with the identification label and then placed in a secured area of a freezer. Alternatively, the bagged sample can be placed in a thin walled forming device (e.g., a paper milk carton) to form the sample to shape the shipping container. When frozen solidly, the sample can be placed in an insulated shipping container with coolant canisters and sealed.*

TABLE 1

Commodity	Instructions for taking a primary sample	Minimum quantity required for laboratory sample
I. Group 030 (Mammalian Meats)		
A. Whole carcass or side, unit weight normally 10 kg or more	Take diaphram muscle supplemented by cervical muscle, if necessary, from one animal.	0.5 kg.
B. Small carcass (e.g., rabbit)	Take hind quarters or whole carcasses from one or more animals to meet laboratory sample size requirements.	0.5 kg after removal of skin and bone.
C. Fresh/chilled parts		
1. units weighing at least 0.5 kg, excluding any bone, (e.g., quarters, shoulders, roasts)	Take muscle portion from one unit.	0.5 kg.
2. units weighing less than 0.5 kg, (e.g. chops, fillets)	Take number of units from selected container to meet laboratory sample size requirement.	0.5 kg after removal of any bone.
D. Bulk frozen parts	Take a frozen cross-section from selected container or, alternatively, take muscle from one large part.	0.5 kg.
E. Retail packaged frozen/chilled parts, or individually wrapped units for wholesale.	For large cuts, take muscle portion from one unit. Otherwise, take appropriate number of units to meet laboratory sample size requirement.	0.5 kg after removal of any bone.
Ia. Group 030 (Mammalian Meats where MRL is expressed in the carcass fat)		
A. Animals sampled at slaughter	See instructions under II., Group 031.	
B. Other meat parts	Trim off 0.5 kg of visible fat, or take sufficient product to yield 50-100 g of fat for analysis. (Normally 1.5-2.0 kg is required for cuts without trimmable fat.)	Sufficient to yield 50-100 g of fat.

TABLE 1 (continued)

Commodity	Instructions for taking a primary sample	Minimum quantity required for laboratory sample
II. Group 031 (Mammalian fat)		
A. Large animals sampled at slaughter, usually weighing at least 10 kg.	Take kidney, abdominal or subcutaneous fat from one animal.	0.5 kg.
B. Small animals sampled at slaughter*	Take abdominal and sub-cutaneous fat from one or more animals.	0.5 kg.
C. Bulk fat tissue	Take equal size portions from 3 locations in container.	0.5 kg.
III. Group 032 (Mammalian Edible Offal)		
Liver	Take whole liver(s) or portion sufficient to meet laboratory sample size requirement.	0.4-0.5 kg.
Kidney	Take one or both kidneys, or kidneys from more than one animal sufficient to meet laboratory sample size requirement. Do not collect from more than one animal if size meets the low range for the laboratory sample size requirement.	0.25-0.5 kg.
Heart	Take whole heart or ventricle portion sufficient to meet laboratory sample size requirement.	0.4-0.5 kg
Other fresh/chilled or frozen, edible offal product	Take portion derived from one animal unless product from more than one animal is required to meet laboratory sample size requirement. A cross-section can be taken from bulk-frozen product.	0.5 kg.

* *When adhering fat is insufficient to provide a suitable sample, the whole commodity, without bone, is analyzed and the MRL applies to the whole commodity.*

TABLE 1 (continued)

Commodity	Instructions for taking a primary sample	Minimum quantity required for laboratory sample
IV. Group 036 (Poultry Meats)		
A. Whole carcass of large bird, typically weighing 2-3 kg or more (e.g., turkey, mature chicken, goose, duck)	Take thighs, legs, and other dark meat from one bird.	0.5 kg after removal of skin and bone.
B. Whole carcass of bird, typically weighing between 0.5 and 2 kg. (e.g., young chicken, duckling, guinea fowl)	Take thighs, legs and other dark meat from 3 to 6 birds, depending on size.	0.5 kg after removal of skin and bone.
C. Whole carcasses of very small birds typically weighing less than 0.5 kg (e.g. quail, pigeon)	Take at least 6 whole carcasses.	0.25-0.5 kg of muscle tissue.
D. Fresh/chilled or frozen parts.		
1. Wholesale packaged		
a. large parts	Take an interior unit from selected container.	0.5 kg after removal of skin and bone.
b. small parts	Take sufficient parts from a selected layer in the container.	
2. Retail packaged	Take number of units from selected container to meet laboratory sample size requirement.	0.5 kg after removal of skin and bone.
IVa. Group 036 (Poultry meats where MRL is expressed in the carcass fat)		
A. Birds sampled at slaughter	See instructions under V., Grp 037.	
B. Other poultry meat	Take 0.5 kg of separable fat or sufficient product to yield 50-100 g of fat. (Normally, 1.5 - 2 kg is required if separable fat is not available)	Sufficient to yield 50-100 g of fat.

TABLE 1 (continued)

Commodity	Instructions for taking a primary sample	Minimum quantity required for laboratory sample
V. Group 037 (Poultry Fats)		
A. Birds sampled at slaughter	Take abdominal fat from 3 to 6 birds depending on size.	Sufficient to yield 50-100 g of fat.
B. Bulk fat tissue	Take equal size portions from 3 locations in container.	0.5 kg.
VI. Group 038 (Poultry Edible Offal)		
A. Liver	Take 6 whole livers or portion sufficient to meet laboratory sample size requirement.	0.25 - 0.5 kg.
B. Other fresh/chilled or frozen edible offal product	Take appropriate parts from 6 birds; if bulk frozen, take a cross-section from selected container.	0.25 - 0.5 kg.
VII. Class E - Type 16 (Secondary Meat and Poultry Products)		
A. Fresh/chilled or frozen comminuted product of single species origin	Take a representative or fresh cross section from selected container or packaged unit.	0.5 kg.
B. Group 080 (Dried Meat Product)	Take number of packaged units in a selected container sufficient to meet laboratory sample size requirements.	0.5 kg, unless fat content is less than 5% and MRL is expressed on a fat basis. In this case, 1.5 - 2 kg is required.

TABLE 1 (continued)

Commodity	Instructions for taking a primary sample	Maximum quantity required for laboratory sample
VIII. <u>Class E - Type 18</u> (Manufactured, single ingredient product of meat or poultry origin) A. Canned product, (e.g. ham, beef, chicken - unit size of at least 1 kg)	Take one can from a lot. When unit size is very large, (< 2 kg) a representative sample including liquids may be taken.	0.5 kg, unless fat content is less than 5% and MRL is expressed on a fat basis. In that case, 1.5 - 2 kg is required.
B. Cured, smoked, or cooked product (e.g. bacon slab, ham, turkey, cooked beef - unit size of at least 1 kg).	Take portion from a large unit (> 2 kg), or take whole unit, depending on size.	0.5 kg, unless fat content is less than 5% and MRL is expressed on a fat basis. In that case, 1.5 - 2 kg is required.
IX. <u>Class E - Type 19</u> (Manufactured, multiple ingredient, product of meat and poultry origin) A. Sausage and luncheon meat rolls - unit size of at least 1 kg.	Take a cross section portion from a large unit (> 2 kg) or whole unit, depending on size.	0.5 kg.

SECTION 4

ANALYSIS OF PESTICIDE RESIDUES

SECTION 4.1

PORTION OF COMMODITIES TO WHICH
CODEX MRLS APPLY
AND WHICH IS ANALYZED

PORTION OF COMMODITIES
TO WHICH CODEX MAXIMUM RESIDUE LIMITS APPLY
AND WHICH IS ANALYZED

INTRODUCTION

Codex maximum residue limits are in most cases stated in terms of a specific whole raw agricultural commodity as it moves in international trade. In some instances, a qualification is included that describes the part of the raw agricultural commodity to which the maximum residue limit applies, for example, almonds on a shell-free basis and beans without pods. In other instances, such qualifications are not provided. Therefore, unless otherwise specified, the portion of the raw agricultural commodity to which the MRL applies and which is to be prepared as the analytical sample for the determination of pesticide residues is as described in the following table.

CLASSIFICATION OF COMMODITIES

PORTION OF COMMODITY TO WHICH THE CODEX MRL APPLIES (AND WHICH IS ANALYZED)

Group 1 - ROOT AND TUBER VEGETABLES

Root and tuber vegetables are starchy foods derived from the enlarged solid roots, tubers, corms or rhizomes, mostly subterranean, of various species of plants. The entire vegetable may be consumed.

Root and Tuber vegetables:

beets	radishes
carrots	rutabagas
celeriac	sugar beets
parsnips	sweet potatoes
potatoes	turnips
	yams

Whole commodity after removing tops. Remove adhering soil (e.g. by rinsing in running water or by gentle brushing of the dry commodity)

CLASSIFICATION OF COMMODITIES

PORTION OF COMMODITY TO WHICH THE CODEX MRL APPLIES (AND WHICH IS ANALYZED)

Group 2 - BULB VEGETABLES

Bulb vegetables are pungent flavourful foods derived from the fleshy scale bulbs, or growth buds of alliums of the lily family *(Liliaceae)*. The entire bulb may be consumed following removal of the parchment like skin.

Remove adhering soil (e.g. by rinsing in running water or by gentle brushing of the dry commodity)

Bulb vegetables:

| garlic | onions |
| leeks | spring onions |

Bulb/dry onions and garlic:
Whole commodity after removal of roots and whatever parchment skin is easily detached.
Leeks and spring onions:
Whole vegetable after removal of roots and adhering soil.

Group 3 - LEAFY VEGETABLES (EXCEPT BRASSICA VEGETABLES)

Leafy vegetables (except Group 4 vegetables) are foods derived from the leaves of a wide variety of edible plants including leafy parts of Group 1 vegetables. The entire leaf may be consumed. Leafy vegetables of the brassica family are grouped separately.

Leafy vegetables:

beet leaves	radish leaves
corn salad	spinach
endive	sugar beet leaves
lettuce	Swiss chard

Whole commodity after removal of obviously decomposed or withered leaves.

CLASSIFICATION OF COMMODITIES

PORTION OF COMMODITY TO WHICH THE CODEX MRL APPLIES (AND WHICH IS ANALYZED)

Group 4 - BRASSICA (COLE) LEAFY VEGETABLES

Brassica (cole) leafy vegetables are foods derived from the leafy parts, stems and immature influorescenses of plants commonly known and botanically classified as brassicas and also known as cole vegetables. The entire vegetable may be consumed.

Brassica leafy vegetables:

broccoli	cauliflower
cabbage	collards
cabbage, Chinese	kales
cabbage, red	kohlrabi
cabbage, Savoy	mustard greens

Whole commodity after removal of obviously decomposed or withered leaves. For cauliflower and headed broccoli analyse Brussels sprouts flower head and stems discarding leaves; for Brussels sprouts analyse "buttons" only.

Group 5 - STEM VEGETABLES

Stem vegetables are foods derived from the edible stems or shoots from a variety of plants.

Stem vegetables:

artichoke	chicory (witloof)
celery	rhubarb

Whole commodity after removal of obviously decomposed or withered leaves.
Rhubarb and asparagus: stems only.
Celery and asparagus: remove adhering soil (e.g., by rinsing in running water or by gentle brushing of the dry commodity).

CLASSIFICATION OF COMMODITIES   **PORTION OF COMMODITY TO WHICH THE CODEX MRL APPLIES (AND WHICH IS ANALYZED)**

Group 6 - LEGUME VEGETABLES

Legume vegetables are derived from the dried or succulent seeds and immature pods or leguminous plants commonly known as beans and peas. Succulent forms may be consumed as whole pods or as the shelled product. Legume fodder is in Group 18.

Legume vegetables: Whole commodity.

beans	navy beans
broad beans	runner beans
dwarf beans	snapbeans
French beans	soybeans
green beans	peas
kidney beans	cow peas
Lima beans	sugar peas

Group 7 - FRUITING VEGETABLES - EDIBLE PEEL

Fruiting vegetables - edible peel are derived from the immature or mature fruits of various plants, usually annual vines or bushes. The entire fruiting vegetables may be consumed.

Fruiting vegetables - edible peel   Whole commodity after removal of stems.

cucumber	pepper
egg plant	summer squash
gherkin	tomato
okra	

CLASSIFICATION OF COMMODITIES	PORTION OF COMMODITY TO WHICH THE CODEX MRL APPLIES (AND WHICH IS ANALYZED)

Group 8 - FRUITING VEGETABLES - INEDIBLE PEEL

Fruiting vegetables - inedible peel are derived from the immature or mature fruits of various plants, usually annual vines or bushes. Edible portion is protected by skin, peel or husk which is removed or discarded before consumption.

Fruiting vegetables - inedible peel   Whole commodity after removal of stems.

cantaloupe	squash
melon	watermelon
pumpkin	winter squash

Group 9 - CITRUS FRUITS

Citrus fruits are produced by trees of the rue family and characterized by aromatic oily peels, globular form, and interior segments of juice filled vesicles. The fruit is fully exposed to pesticides during the growing season. The fruit pulp may be consumed in succulent form and as a beverage. The entire fruit may be used for preserving.

Citrus fruits: Whole commodity.

Group 10 - POME FRUITS

Pome fruits are produced by trees related to the *genus pyrus* of the rose family (*Rosaceae*). They are characterized by fleshy tissue surrounding a core consisting of parchment like carpels enclosing the seed. The entire fruit, excepting the core, may be consumed in the succulent form or after processing.

Pome fruits: Whole commodity after removal of stems.

apple	quince
pear	

CLASSIFICATION OF COMMODITIES	PORTION OF COMMODITY TO WHICH THE CODEX MRL APPLIES (AND WHICH IS ANALYZED)

Group 11 - STONE FRUITS

Stone fruits are produced by trees related to
the *genus prunus* of the rose family (*Rosaceae*)
characterized by fleshy tissue surrounding a
single hard shelled seed. The entire fruit,
except seed, may be consumed in a succulent
or processed form.

Stone fruits:

apricots	nectarines
cherries	peaches
sour cherries	plums
sweet cherries	

Whole commodity after removal of stems and
stones but the residue calculated and expressed
on the whole commodity without stem.

Group 12 - SMALL FRUITS AND BERRIES

Small fruits and berries are derived from a
variety of plants having fruit characterized
by a high surface-weight ratio. The entire
fruit, often including seed, may be consumed
in a succulent or processed form.

Small fruits and berries:

blackberries	gooseberries
blueberries	grapes
boysenberries	loganberries
cranberries	raspberries
currants	strawberries
dewberries	

Whole commodity after removal of caps
and stems. Currants: fruit with stems.

CLASSIFICATION OF COMMODITIES	PORTION OF COMMODITY TO WHICH THE CODEX MRL APPLIES (AND WHICH IS ANALYZED)

Group 13 - ASSORTED FRUITS - EDIBLE PEEL

Assorted fruits - edible peel are derived from the immature or mature fruits of a variety of plants, usually shrubs or trees from tropical or subtropical regions. The whole fruit may be consumed in a succulent or processed form.

Assorted fruits - edible peel:

dates
figs
olives

Dates and olives: whole commodity after removal of stems and stones but residue calculated and expressed on the whole fruit.
Figs: Whole commodity.

Group 14 - ASSORTED FRUITS - INEDIBLE PEEL

Assorted fruits - inedible peel are derived from the immature or mature fruits of different kinds of plants, usually shrubs or trees from tropical or subtropical regions. Edible portion is protected by skin, peel or husk. Fruit may be consumed in a fresh or processed form.

Assorted fruits - inedible peel:

avocados	mangoes
bananas	papayas
guavas	passion fruits
kiwi fruit	pineapples

Whole commodity unless qualified.
Pineapples: after removal of crown.
Avocado and mangoes: whole commodity after removal of stone but calculated on whole fruit.
Bananas: after removal of crown tissues and stalks.

CLASSIFICATION OF COMMODITIES

PORTION OF COMMODITY TO WHICH THE CODEX MRL APPLIES (AND WHICH IS ANALYZED)

Group 15 - CEREAL GRAINS

Cereal grains are derived from the clusters of starchy seeds produced by a variety of plants primarily of the grass family (*Gramineae*). Husks are removed before consumption.

Cereal grains:

barley   rye
maize   sorghum
oats   sweet corn
rice   wheat

Whole commodity.
Fresh corn and sweet corn: kernels plus cob without husk.

Group 16 - STALK AND STEM CROPS

Stalk and stem crops are various kinds of plants, mostly of the grass family (*Gramineae*) cultivated extensively as animal feed and for the production of sugar. Stems and stalks used for animal feeds are consumed as succulent forage, silage, or as dried fodder or hay. Sugar crops are processed.

Stalk and stem crops:

barley fodder   maize fodder
  and straw   sorghum fodder
grass fodders

Whole commodity.

Group 17 - LEGUME OILSEEDS

Legume oilseeds are mature seeds from legumes cultivated for processing into edible vegetable oil or for direct use as human food.

Legume oilseeds:

peanuts

Whole kernel after removal of shell.

CLASSIFICATION OF COMMODITIES

PORTION OF COMMODITY TO WHICH THE CODEX MRL APPLIES (AND WHICH IS ANALYZED)

Group 18 - LEGUME ANIMAL FEEDS

Legume animal feeds are various species of legumes used for animal forage, grazing, fodder, hay or silage with or without seed. Legume animal feeds are consumed as succulent forage or as dried fodder or hay.

Legume and animal feeds: Whole commodity.

alfalfa fodder	peanut fodder
bean fodder	pea fodder
clover fodder	soybean fodder

Group 19 - TREE NUTS

Tree nuts are the seed of a variety of trees and shrubs which are characterized by a hard inedible shell enclosing an oil seed. The edible portion of the nut is consumed in succulent, dried or processed forms.

Tree nuts: Whole commodity after removal of shell.
Chestnuts: whole in skin.

almonds	macadamia nuts
chestnuts	pecans
filberts	walnuts

CLASSIFICATION OF COMMODITIES **PORTION OF COMMODITY TO WHICH THE CODEX MRL APPLIES (AND WHICH IS ANALYZED)**

Group 20 - OILSEED

Oilseed consists of the seed from a variety of plants used in the production of edible vegetable oils. Some important vegetable oilseeds are by-products of fibre or fruit crops.

Oilseed: Whole commodity.

cottonseed   safflowerseed
linseed   sunflowerseed
rapeseed

Group 21 - TROPICAL SEEDS

Tropical seeds consist of the seed from several tropical and semitropical trees and shrubs mostly used in the production of beverages and confections. Tropical seeds are consumed after processing.

Tropical seeds: Whole commodity.

cacao beans
coffee beans

Group 22 - HERBS

Herbs consist of leaves, stems and roots from a variety of herbaceous plants used in relatively small amounts to flavour other foods. They are consumed in succulent or dried forms as components of other foods.

Herbs: Whole commodity.

CLASSIFICATION OF COMMODITIES	PORTION OF COMMODITY TO WHICH THE CODEX MRL APPLIES (AND WHICH IS ANALYZED)

Group 23 - SPICES

Spices consist of aromatic seeds, roots, fruits and berries from a variety of plants used in relatively small amount to flavour other foods. They are consumed primarily in the dried form as components of other foods.

Spices: Whole commodity.

Group 24 - TEAS

Teas are derived from the leaves of several plants, but principally *Camellia sinensis*. They are used in the preparation of infusions for consumption as stimulating beverages. They are consumed as extracts of the dried or processed product.

Teas: Whole commodity.

Group 25 - MEATS

Meats are the muscular tissue, including adhering fatty tissue from animal carcasses prepared for wholesale distribution. The entire product may be consumed.

Meats: Whole commodity. (For fat soluble pesticides a portion of carcase fat is analyzed and MRLs apply to carcase fat)[1]

carcase meat (and carcase fat)
carcase meat of cattle
carcase meat of goats
carcase meat of horses
carcase meat of pigs
carcase meat of sheep

---

[1]   *For milk and milk products regarding fat soluble pesticides see Section 1 of this Volume.*

CLASSIFICATION OF COMMODITIES   **PORTION OF COMMODITY TO WHICH THE CODEX MRL APPLIES (AND WHICH IS ANALYZED)**

Group 26 - ANIMAL FATS

Animal fats are the rendered or extracted fat from the fatty tissue of animals. The entire product may be consumed.

Animal fats: Whole commodity

cattle fat
pig fat
sheep fat

Group 27 - MEAT BYPRODUCTS

Meat byproducts are edible tissues and organs, other than meat and animal fat, from slaughtered animals as prepared for wholesale distribution. Examples: liver, kidney, tongue, heart. The entire product may be consumed.

Meat byproducts (such as liver, kidney, etc.):   Whole commodity.

cattle meat byproducts
goat meat byproducts
pig meat byproducts
sheep meat byproducts

Group 28 - MILKS

Milks are the mammary secretion of various species of lactating herbivorous ruminant animals, usually domesticated. The entire product may be consumed.

Milks: [2]   Whole commodity.

---

[2]   *For milk and milk products regarding fat soluble pesticides see Section 1 of this Volume.*

CLASSIFICATION OF COMMODITIES	PORTION OF COMMODITY TO WHICH THE CODEX MRL APPLIES (AND WHICH IS ANALYZED)

Group 29 - MILK FATS

Milk fats are the rendered or
extracted fats from milk.

Milk fats: Whole commodity.

Group 30 - POULTRY MEATS

Poultry meats are the muscular tissues including
adhering fat and skin from poultry carcasses
as prepared for wholesale distribution. The
entire product may be consumed.

Poultry Meats: Whole commodity. (For fat soluble pesticides
a portion of carcase fat is analyzed and MRLs
apply to carcase fat).

Group 31 - POULTRY FATS

Poultry fats are the rendered or extracted fats
from fatty tissues of poultry. The entire
product may be consumed.

Poultry fats: Whole commodity.

Group 32 - POULTRY BYPRODUCTS

Poultry byproducts are edible tissue and organs,
other than poultry meat and poultry fat from
slaughtered poultry.

Poultry byproducts: Whole commodity.

CLASSIFICATION OF COMMODITIES   **PORTION OF COMMODITY TO WHICH THE CODEX MRL APPLIES (AND WHICH IS ANALYZED)**

Group 33 - EGGS

Eggs are the fresh edible portion of the reproductive body of several avian species. The edible portion includes egg white and egg yolk after removal of the shell.

Eggs: Whole egg whites and yolks combined after removal of shells.

SECTION 4.2

GUIDELINES ON GOOD LABORATORY PRACTICE
IN
PESTICIDE RESIDUE ANALYSIS

GUIDELINES ON GOOD LABORATORY PRACTICE
IN PESTICIDE RESIDUE ANALYSIS

INTRODUCTION

The ultimate goal in fair practice in international trade depends, among other things, on the reliability of analytical results. This in turn, particularly in pesticide residues analysis, depends not only on the availability of reliable analytical methods, but also on the experience of the analyst and on the maintenance of "good laboratory practice in the analysis of pesticides". These guidelines define such good analytical practice and may be considered in three inter-related parts:

The Analyst;
Basic Resources, and
The Analysis.

A discussion of each of these follows:

THE ANALYST

Residue analysis consists of a chain of procedures, most of which are known, or readily understood, by a trained chemist, but because the acceptable margin for error is smaller than in most types of analysis and any mistake can invalidate the whole analysis, attention to detail is essential. The analyst in charge should have an appropriate professional qualification and be experienced and competent in residue analysis. At any one time the majority of staff need to be trained in the correct use of apparatus and basic laboratory skills and the principles of residue analysis. They must understand the purpose of each stage in the method being used and the importance of following the method exactly as described and of noting any unavoidable deviations. A clear understanding of the terminology involved is also essential.

Ideally, when a laboratory for residue analysis is set up, the staff should spend some of their training period in a well established laboratory where experienced advice and training is available. If the laboratory is to be involved in the analysis for a wide range of pesticide residues it may be necessary for the staff to gain experience in more than one established laboratory.

BASIC RESOURCES

The Laboratory

In ideal circumstances the laboratory and its fittings should be designed to allow tasks to be allocated to well defined areas where maximum safety and minimum chance of contamination of samples prevail. Fittings should be of materials resistant to attack by chemicals likely to be used in the area. Thus, under such ideal conditions, separate rooms would be designated for sample receipt and storage, for sample preparation, for extraction and clean-up and for instrumentation used in the determinative step.

The area used for extraction and clean-up would meet solvent laboratory specifications and all fume extraction facilities would be of high quality. The minimum requirements for pesticide residue analysis are that the facilities are adequate to avoid contamination.

Laboratory safety must also be considered in terms of necessary and preferable conditions as it must be recognized that the stringent working conditions enforced in residue laboratories in some parts of the world would be totally unrealistic in others. No smoking, eating, drinking or application of cosmetics should be permitted in the working area. Only small volumes of solvents should be held in the working area and the bulk of the solvents stored separately, away from the main working area. The use of highly or chronically toxic solvents and reagents should be avoided whenever possible. All waste solvent should be stored safely and disposed of frequently.

The main working area should be treated as a solvent laboratory and all equipment such as lights, macerators and refrigerators should be "spark free" or "explosion proof". Extractions, clean-up and concentration steps should be carried out in a well ventilated area, preferably in fume cupboards. Fume hoods or canopies above the bench are generally ineffective in removing vapours at the working level.

Safety screens should be used when glassware is used under vacuum or pressure. There should be an ample supply of safety glasses, gloves and other protective clothing, emergency washing facilities and spillage treatment kits. Adequate emergency fire extinguishing equipment must be provided. All staff should be trained in the use of these facilities and in an appreciation of the hazards involved. Staff must be aware that many pesticides have acute or chronically toxic properties and although little risk is attached to the handling of most samples, great care is necessary in the handling of standard reference compounds.

The staff should be given periodic medical checks.

Equipment and Supplies

The laboratory will require adequate supplies of electricity and water and various gases, either piped or from gas cylinders, of proven quality. Adequate supplies of reagents, solvents, glassware, stationary phases, etc., are essential.

Servicing facilities for gas chromatographs, balances, spectophotometers, etc., will be required and will probably involve keeping some essential spare parts and access to a good technical service.

Although, in an ideal situation, equipment should be regularly updated in order to keep up with developments, e.g. gas chromatography with microprocessor controls, the equipment only needs to be sophisticated enough to do the job required. Thus, the demands for monitoring commodities at specified Codex Maximum Residue Limits (MRLs) may be much less stringent than those required in a research environment.

All laboratories require an adequate range of reference standard pesticides of known and reasonably high purity. The range should cover all parent species for which the laboratory is monitoring samples as well as those metabolites which are included in MRLs.

THE ANALYSIS

Avoidance of Contamination

One of the major areas in which pesticide residue analysis differs significantly from macro-analysis is that of the problem of contamination. Trace amounts of contamination in the final samples used for the determination stage of the method can give rise to errors such as false positive results and to a loss of sensitivity that may prevent the residue analyst from achieving the necessary limits of determination. Contamination may arise from either construction material, from the environment or from the procedure.

Bench polish, barriers creams, soaps containing germicides, fly sprays, perfumes and cosmetics are all commodities that give rise to laboratory contamination and are especially significant when an electron-capture detector is being used. There is no real solution to the problem other than to ban their use in the laboratory.

Greases, plasticizers, rubber bungs and tubing, oil from air lines, extraction thimbles, filter-papers and cotton-wool can also give rise to contamination of the final test solution.

Pesticide reference standards should always be stored in a room separate from the main residue laboratory. Field samples, sample preparation and formulation analysis should also be kept separate from the main residue laboratory.

Contamination of glassware, syringes and gas chromatographic columns can arise from previous samples. All glassware should be cleaned with detergent, rinsed thoroughly and then rinsed with the solvent to be used. There must be a separate stock of glassware for pesticide residue work.

Chemical reagents, adsorbents and general laboratory solvents may contain components that interfere in the analysis. It may be necessary to purify reagents and adsorbents by heating and it is generally necessary to use redistilled solvents. De-ionized water is often suspect and redistilled water is preferable. In many instances tap water or well water may be satisfactory. All glassware, reagents, organic solvents and water should be checked for possible contaminants before use.

Apparatus containing polyvinylchloride should be suspect and, if shown to be a source of contamination, should not be allowed in the residue laboratory. Other materials containing plasticizers are suspect but PIFE and silicone rubbers are usually acceptable and others may be acceptable in certain circumstances. Sample storage containers can cause contamination and glass bottles with ground glass stoppers should always be used. Instrumentation should always be housed in a separate room. The nature and importance of contamination can vary according to the type of determination technique used and the level of pesticide residue to be determined. These contamination problems, which are important with methods based on gas chromatography or high performance liquid chromatography, may well be less significant if a spectrophotometric determination is used, and vice-versa. For relatively high levels of residues the background interference from solvents and other materials may be insignificant in comparison with the amount of residue present. Also, many problems can be solved by the use of specific detectors. Furthermore, if the contaminant does not interfere with the residue being sought, its presence may be acceptable.

Avoidance of Losses

In an ideal situation samples should be stored at chill (3-5°C) temperature, away from direct sunlight, and analyzed within a few days. However, in many instances samples may require storage for an extended period (6-9 months) before analysis and the following precautions should be observed:

a. Storage temperature should be approximately -20°C, at which temperature degradation of residues of pesticides by enzyme action is extremely slow. If any doubts exist, the samples should be compared with fortified samples stored under the same conditions.

b. All samples should be re-homogenised after freezing as there is a tendency for water to distil out and to collect as ice crystals, which, if discarded, will affect the analytical result.

c. Neither the containers used for storage nor their caps or stoppers should allow migration of the chemical being sought into the container. The containers must not leak. All samples should be labelled clearly with permanent labels and recorded in a sample book.

The extracts and final test solutions should not be exposed to direct sunlight.

Standard Operating Procedures

Written, adequate, standard operating procedures must be available for use by analytical staff. Any deviations from procedures must be recorded and authorised by the analyst in charge.

Validation of Methods

The amount of effort allocated to the validation of methods will vary considerably. In a routine laboratory monitoring for compliance with Codex MRLs or national tolerances, standardized methods will be used in most instances. These should be validated initially and thereafter checked periodically.

In all laboratories, regular checks should be made on the effects of variation in sources of supply of chemicals, solvents, etc. The performance of the method should be checked by, for example, the recovery of standards, added at appropriate levels, taken through the method both alone and in the presence of each new substrate.

In laboratories where method development and/or modification is undertaken, other aspects that may be studied are the effect of variation in sample size, partition ratios, etc., the efficiency, resolution and column stability of gas- and liquid-chromatographic systems and variations in activity of various column clean-up systems. The effects of light, storage at intermediate stages of the procedure, temperature, etc., on the stability of reagents and samples should also be studied. The evaluation of detection/determination systems (e.g., in gas or liquid chromatography) for effects of flow-rate, temperature, etc., is important.

Maintenance of Overall Analytical Performance

In all laboratories engaged in pesticide residue analysis there is a need for regular assessment of the methods in use, both at the tolerance level and at the lower limit of determination.

Recovery of pesticides from "spiked" samples is commonly used as a measure of efficiency of extraction, but it must be recognized that such studies are of limited value. Where possible, emphasis should be placed on checking recoveries where residues are in a "real" state, e.g. in field treated samples, since a method that gives adequate recoveries from samples spiked with parent compounds may be inadequate for the efficient extraction of the parent compounds and their metabolites from the "real" samples. The evaluation of a method should include, where possible, the extraction of radiolabelled compounds or a comparative extraction with a method of known efficiency. Samples spiked with the parent compound and any known metabolites should also be analyzed to determine the loss of these compounds during analysis. Recoveries should be within the range 70-110% with a mean of greater than 80% after removal of outliers.

Regular analyses of substrates known to be free of pesticide residues is necessary in order to check that contamination is not occurring.

Care should be taken that standard solutions of pesticides are not decomposed by the effect of light or heat during storage or become more concentrated owing to solvent evaporation. Equal care must be taken to ensure the stability of reference standard compounds. Regular injection of standards during gas chromatographic analysis of a series of samples allows the performance of the determinative step to be checked.

A means of monitoring the performance of a method (or an analyst) is to introduce check samples at regular intervals. These check samples should be introduced as routine samples without any indication being given as to their special nature.

Various national and international organizations now organize collaborative studies on particular methods and/or check sample programmes. The latter present an ideal way for laboratories to assess their own performance. If possible, check samples should be introduced as routine samples so that the analyst concerned does not attempt to "make a special effort", which would invalidate the samples as a test of laboratory performance.

Confirmatory Tests

When analyses are done for regulatory purposes it is especially important that confirmatory tests are carried out before reporting adversely on samples containing residues of pesticides not normally associated with that commodity or where MRLs appear to have been exceeded.

Contamination of samples with non-pesticidal chemicals occurs from time to time and in some chromatographic methods these compounds may have similar properties to pesticides and may therefore be mis-identified as such. Examples in gas-chromatography include the responses of electron-capture detectors to phthalate esters and of phosphorus-specific detectors to compounds containing sulfur.

Confirmatory tests can be divided into two types. Quantitative tests are necessary when MRLs appear to be exceeded whilst qualitative confirmation of identity is also needed in these cases, and when atypical residues are encountered. Qualitative tests may involve chemical reactions or separations where some loss of the residue occurs. Particular problems occur in confirmation when MRLs are set at or about the limit of analytical determination.

The need for confirmatory tests may depend upon the type of sample or its known history. In many substrates, certain residues are nearly always found. For a series of samples of similar origin it may only be necessary to confirm the identity of residues in the initial samples. Similarly, when it is known that a particular pesticide has been applied to the sample material there may be little need for confirmation of identity, although a random proportion of samples should be confirmed. Where control samples are available, these should be used to check for the presence of possible interfering substances.

In quantitative confirmation at least one alternative procedure should be used and the lower result reported. In qualitative confirmation, an alternative technique using different physico-chemical properties is desirable.

The necessary steps to positive identification are a matter of judgement on the analyst's part and particular attention should be paid to the choice of a method which would eliminate the effect of interfering compounds. The chosen method would depend upon the availability of suitable apparatus and expertise within the testing laboratory. As guidance to the analyst a number of alternative procedures for confirmation are given in the following paragraphs.

Alternative gas chromatographic columns

The results obtained in the primary analysis should always be quantitatively and qualitatively confirmed using at least one alternative column involving a stationary phase of different polarity. The quantitative results obtained should be within 20% of the primary analysis and the lower figure should be reported, since the higher figure may have been enhanced by interference from co-extracted material. Further quantitative confirmation is required if the results differ by more than 20%, except when the MRL is set "at or about the limit of determination" when a variation of up to 100% would be acceptable. In choosing the alternative column material, consideration should be given to separating any pesticidal or interfering compounds known to have retention times on the primary column identical to that of the residue detected. The alternative column may be a packed column or, preferably, a capillary column whose greater resolving power may be utilized. Whilst the use of an alternative gas-chromatographic column may not always give positive confirmation it will often quickly disprove a suspected identity. In either case further confirmation is required to identify the residue.

Use of selective detectors for gas chromatography

When pesticides containing several chemical elements are present, detectors showing enhanced response to these elements may be used. Detectors such as flame photometric (sulfur, phosphorus and tin), alkali flame ionization (phosphorus and nitrogen) and coulometric/conductivity (nitrogen, sulfur and halogens) can give valuable additional information on residues. The sulfur/phosphorus response ratio obtained by using a flame photometric detector can give useful information in the case of phosphorothioates.

Thin-layer chromatography (TLC)

In some instances, confirmation of gas-chromatographic findings is most conveniently achieved by TLC. Identification is based on two criteria, Rf value and visualization reaction. The scientific literature contains numerous references to the technique, the IUPAC Report on Pesticides (Bátora, V., Vitorovic, S.Y., Thier, H.P. and Klisenko, M.A., Pure and Applied Chemistry 53, 1039-1049, 1981) reviews the technique and serves as a convenient introduction. The quantitative aspects of thin-layer chromatography are, however, limited. A further extension of this technique involves the removal of the area on the plate corresponding to the Rf of the compound of interest following by elution from the layer material and further chemical or physical confirmatory analysis. A solution of the standard pesticide should always be spotted on the plate alongside the sample extract to obviate any problems of non-repeatability of Rf. Overspotting of extract with standard pesticide can also given useful information. The advantages of thin-layer chromatography are speed, low cost and applicability to heat sensitive materials; disadvantages include (usually) lower sensitivity than gas-liquid chromatography and frequent need for more efficient clean-up. In some countries problems may be encountered when high humidity or temperature cause lack of repeatability.

High performance liquid chromatography (HPLC)

HPLC can often be used advantageously for the confirmation of residues initially found by gas chromatography or by other techniques and may be in certain circumstances the preferred quantitative technique. Post- or pre-column derivatization, and/or use of different detectors, are further options available to the analyst, especially when heat sensitivity or low volatility make the compound to be analyzed less amenable to gas chromatography.

Column fractionation

The order of elution from liquid chromatographic columns may help to verify the identity of a compound. Thus, an element of confirmation can be built-in to the extraction and clean-up procedure.

Extraction p-values

The fraction of pesticide partitioning into the upper phase when distributed between equal volumes of two immiscible liquids has been defined as the extraction p-value and is often a unique value for a given pesticide-solvent system.

The technique has been fully described by Beroza, M. and Bowman, M.C. (Anal. Chem. 37, 291, 1965; JAOAC 48, 358, 1965; JAOAC 48, 943, 1965) and Beroza, M., Inscoe, M.N. and Bowman, M.C. (Residue Reviews 30, 1, 1969).

Derivatization

This area of confirmation may be considered under three broad headings:

a.   Chemical reactions

Small scale chemical reactions resulting in degradation, addition or condensation products of pesticides, followed by re-examination of the products by chromatographic techniques, have frequently been used. The reactions result in products possessing different retention times and/or detector response from those of the parent compound. A sample of standard pesticide should be treated alongside the suspected residue so that the results from each may be directly compared. A fortified extract should also be included to prove that the reaction has proceeded in the presence of sample material. A review of chemical reactions which have been used for confirmatory purposes has been published by Cochrane, W.P. (Chemical derivatization in pesticide analysis, Plenum Press, N.Y, 1981). Chemical reactions have the advantages of being fast and easy to carry out, but specialized reagents may need to be purchased and/or purified.

b.   Physical reactions

A useful technique is the photochemical alteration of a pesticide residue to give one or more products with a reproducible chromatographic pattern. A sample of standard pesticide and fortified extract should always be treated in a similar manner. Samples containing more than one pesticide residue may give problems in the interpretation of results. In such cases pre-separation of specific residues may be carried out using TLC, HPLC or column fractionation (see techniques above) prior to reaction.

c.   Other methods

Many pesticides are susceptible to degradation/transformation by enzymes. In contrast to normal chemical reactions, these processes are very specific and generally consist of one of the following: oxidation, hydrolysis or de-alkylation. The products possess different chromatographic characteristics from the parent pesticide and may be used for confirmatory purposes if compared with reaction products using standard pesticides.

Mass Spectrometry

Results obtained using mass spectrometry present the most definitive evidence for confirmation/identification purposes. Where the apparatus is available it is usually the confirmatory technique of choice.

There are two principal methods of introducing samples into the instrument. The preferred method utilizes gas-chromatographic separation prior to introduction into the mass spectrometer. This allows full mass spectral analysis of the peak observed during the primary analysis. Alternatively, samples can be introduced using the direct insertion probe technique. This method can be used in conjunction with TLC or HPLC when these have been used as initial confirmatory procedures. Residues separated into these techniques are isolated and subjected to mass spectrometry.

To increase sensitivity, particularly with fast scanning quadrupole instruments, techniques known as single and multiple ion detection have been used. A sufficient number of fragment ions must be selected to ensure unambiguous identification. Increased sensitivity with respect to the molecular ion may be obtained by using chemical ionization in place of electron-impact. As mass spectrometers are generally sensitive at the nanogram level some extracts from primary gas chromatographic analysis may require concentration before mass spectrometric analysis, particularly when electron-capture detectors have been used for quantification. In some cases additional clean-up may be necessary, particularly if full spectra are to be obtained.

Problems can be encountered with heat-sensitive compounds during mass spectrometry and particular care has to be taken when coupling gas chromatographs to mass spectrometers. As there is almost no differential response to compounds in mass spectrometry, complications can arise in the presence of co-eluting contaminants. A useful introduction of the various techniques is to be found in "Mass Spectrometry of Pesticides and Pollutants" by O. Hutzinger and F. Safe, CRC Press 1973 and "Biochemical Applications of Mass Spectrometry" first supplemental volume, chapter 23 by J.A. Sphon and W.C. Brumley: Book edited by G.R. Waller and O.C. Dormer, John Wiley & Sons, New York 1980.

Spectral measurements

At present little use is made of infra-red, Raman or nuclear magnetic resonance spectroscopy in pesticide residue analysis. Instrumental techniques using multiple reflection cells, microcells, microphobes, laser light, Fourier Transformation etc. are being developed. These improve the quality of spectra and enhance the sensitivity and may enlarge the application of these techniques as post-column detection methods for identification of compounds isolated by chromatographic techniques.

SECTION 4.3

RECOMMENDED METHODS OF ANALYSIS
OF
PESTICIDE RESIDUES

RECOMMENDED METHODS OF ANALYSIS
OF PESTICIDE RESIDUES

INTRODUCTION

SCOPE

In this Section are recommendations for analytical methods which can, from practical experience of the Working Group on Methods of Analysis to the Codex Committee on Pesticide Residues (CCPR), be applied to the determination of pesticide residues for regulatory purposes. The "List of Methods of Analysis", to follow, is not exhaustive and methods not mentioned in the List can also be applied under certain conditions. Following the List are "References to Literature", which include references to general articles, laboratory manuals and individual papers in the scientific literature, all of which deal with pesticide residue methodology.

CRITERIA FOR THE SELECTION OF ANALYTICAL METHODS

Whenever possible, the following criteria were used when selecting analytical methods for inclusion in the List:

a.   published in books, manuals or open literature; (For some newer compounds, few methods might be available from these sources; in those cases, GIFAP is prepared to supply analytical methods to bona fide scientists. Requests can be directed to: GIFAP, Avenue Albert Lancaster 79A, 1180 Brussels, Belgium);

b.   collaboratively studied or known to have been validated in a large number of laboratories;

c.   capable of determining more than one residue, i.e. multi-residue methods;

d.   suitable for as many commodities as possible at or below the specified MRLs;

e.   applicable in a regulatory laboratory equipped with routine analytical instrumentation.

Preference was given to gas chromatography or high performance liquid chromatography as the determinative step for the recommended methods. Under certain conditions however, methods using less sophisticated procedures, such as thin-layer chromatography or spectrophotometry, may be applicable. This may be the case, for example, when an exporting country wants to check whether or not a commodity produced in that country complies with an Codex MRL. In this case, the treatment history of the commodity may be known or assumed, so that the method used need not be as elaborate as in cases where samples of unknown treatment history are under investigation. Also, when the MRL is high compared to the limit of determination, simpler methodology may be applied in order to arrive at a "pass/no pass" decision or for quick screening purposes.

APPLICATION OF METHODS

It will always be necessary for the analyst to validate a method before it is first applied in a practical situation. There is a further need for regular checks on the performance of the method in use at both the MRL and at the lower limit of determination. The methods are only recommended for the pesticide/commodity combinations reported in the quoted references. For all new pesticide/commodity combinations the method must be validated following Good Laboratory Practice in Residue Analysis, (see Section 4.2 of this Volume). Confirmation of the identity of an indicated residue by an independent technique is also to be regarded as an essential part of Good Laboratory Practice in Residue Analysis, especially when the initial result suggests that an MRL is exceeded. Mass spectrometry has become for many residues the method of choice for confimatory purposes, but the ultimate choice of a confirmatory test depends upon the technique used in the initial determination and upon the available instrumentation and necessary expertise.

LIST OF METHODS OF ANALYSIS

In the following List, the CCPR number and compound names are those assigned by Codex to identify the pesticide chemicals and are used, among others, in listings of recommended Codex MRLs for residues of those compounds (see Section 1 of this Volume).

The "References to Analytical Methods" in the List have both numbers and names. They refer to entries in the succeeding subsection entitled "References to Literature". The **numbers** correspond to references under "Manuals" and the **names** to the (first) author of the scientific articles listed in "Individual Papers".

CCPR number	Compounds	References to Analytical Methods (see "References to Literature" to follow)
001	aldrin/dieldrin	1a, 1n, 1o, 1p, 2a, 2d, 2f, 3, 4 (XII-5,6; S1-5, S8-10, S12, S19), 5, 7a (5,6), 7c (S8-10, S12, S19), 8a, 8b, 8c, 8d, 9a (M1, M12), 10 Ambrus, Abbott (2), Panel (4), Stijve (2,3)
002	azinphos-methyl	2c, 2d, 2e, 2f, 3, 4 (XII-6; S5, S8, S19; 63, 63A), 7a (6), 7c (S8, S19), 9a (M2, M5, M12), 10 Abbott (1), Ambrus, Panel (3)
003	binapacryl	2a, 2d, 3, 4 (XII-4,6; S19; 8, 43), 7a (6), 7c (S19), 9b, 10 Baker, PB (2)
004	bromophos	2a, 2c, 2d, 4 (XII-3,6; S5, S8-10, S13, S17, S19; 210, 210A), 6d, 7a (3,6), 7c (S8-10, S13, S17, S19), 9a (M2, M5, M12), 10 Abbott (1), Ambrus, Bottomley, Panel (7,8), Stijve (7)
005	bromophos-ethyl	2a, 2c, 2d, 3, 4 (XII-3,6; S8, S13, S17, S19; 263), 6d, 7a (3,6), 7c (S13, S17, S19), 9a (M2, M5, M12), 10 Abbott (1), Ambrus
006	captafol	2d, 2e, 4 (XII-6; S8, S19, S20; 266, 266A), 6d, 7a (6), 7b, 7c (S8, S19, S20), 9a (M1, M12), 10 Ambrus, Baker, PB (1), Buettler, Gilvydis, Pomerantz

CCPR number	Compounds	References to Analytical Methods (see "References to Literature" to follow)
007	captan	1g, 2a, 2d, 2e, 3, 4 (XII-6; S8, S12, S19, S20; 12, 12A), 7a (6), 7b, 7c (S8, S12, S19, S20), 9a (M1, M12), 10 Ambrus, Baker, PB (1), Buettler, Gilvydis, Pomerantz
008	carbaryl	1e, 1h, 1q, 2d, 2e, 2f, 2g, 3, 4 (XII-6; 100), 6c, 7a (6), 9a (M2, M13), 10 Brauckhoff, Chaput, Lawrence(1)
009	carbon disulphide	9a (M8) Mestres (2)
010	carbon tetrachloride	1d, 9a (M8) Daft, Mestres (2), Panel (5)
011	carbophenothion	1c, 2a, 2c, 2d, 2e, 2f, 3, 3d, 4 (XII-5,6; S8, S10, S13, S16, S19), 7a (5,6), 7c (S8, S10, S13, S16, S19), 8b, 8e, 9a (M2, M5, M12), 10 Abbott (1), Ambrus
012	chlordane	1a, 1o, 2a, 2d, 2f, 3, 4 (XII-5,6; S9, S10, S12, S19), 5, 7a (5,6), 7c (S9, S10, S12, S19), 6c, 6d, 8a, 8b, 8c, 8d, 9a (M1, M12), 10 Panel (4), Stijve (3), Veierov
013	chlordimeform	2e, 6a, 9a (M4), 10
014	chlorfenvinphos	2c, 2d, 2e, 2f, 3, 4 (XII-3,5,6; S8, S13, S17, S19; 239), 5, 7a (3,5,6), 7c (S8, S13, S17, S19), 9a (M2, M5, M12), 10 Abbott (1), Ambrus, Panel (7,8), Stijve (7)
015	chlormequat	6a, 9b Sachse, Stijve (5)
016	chlorobenzilate	2a, 2d, 2e, 3, 4 (XII-6; S19), 7a (6), 7c (S19), 10

CCPR number	Compounds	References to Analytical Methods (see "References to Literature" to follow)
017	chlorpyrifos	1p, 2a, 2c, 2d, 2e, 2f, 3, 4 (XII-6; S8, S9, S13, S19), 5, 7a (6), 7c (S8, S9, S13, S19), 8b, 8e, 9a (M2, M5, M12), 10 Ambrus, Stijve (7)
018	coumaphos	2c, 2d, 2e, 3, 4 (XII-6; S19), 7a (6), 7c (S19), 8b, 8e, 9a (M2, M5, M12) Ambrus, Stijve (7)
019	crufomate	2d, 2e, 2f, 4 (XII-6; S19), 7a (6), 7c (S19), 8b, 8e Stijve (7)
020	2,4-D	2b, 2f, 3, 4 (27, 27A), 5, 9a (M6) Ebing, Specht (1)
021	DDT	1a, 1n, 1o, 1p, 2a, 2d, 2f, 3, 4 (XII-4,5,6; S1-5, S8-10, S12, S19), 5, 6c, 7a (4,5,6), 7c (S8-10, S12, S19), 8a, 8b, 8c, 9a (M1, M12), 10 Abbott (2), Ambrus, Bottomley, Panel (4), Stijve (2,3), Veierov
022	diazinon	1a, 2a, 2c, 2d, 2f, 3, 4 (XII-5,6; S5, S8, S10, S13, S17, S19; 35A, 35B), 6c, 7a (5,6), 7c (S8, S10, S13, S17, S19), 8e, 9a (M2, M5, M12), 10 Abbott (1), Ambrus, Bottomley, Panel (7), Stijve (7)
023	1,2-dibromoethane	1d, 8f, 9a (M8) Daft, Heikes, Mestres (2), Panel (5), Rains
024	1,2-dichloroethane	1d, 9a (M8) Daft, Mestres (2), Panel (5)
025	dichlorvos	2c, 2d, 2e, 2f, 3, 4 (XII-3,6; S5, S8, S13, S17, S19; 200), 7a (3,6), 7c (S13, S17, S19), 8b, 8e, 9a (M2, M5), 10 Abbott (1), Ambrus, Bottomley, Panel (1,3,7), Stijve (7)

CCPR number	Compounds	References to Analytical Methods (see "References to Literature" to follow)
026	dicofol	2a, 2d, 2f, 3, 4 (XII-6; S8, S9, S12, S19; 69, 69A), 7a (6), 7c (S8, S9, S12, S19), 9a (M1, M12), 10
027	dimethoate	2c, 2d, 2f, 3, 4 (XII-3,6; S5, S8, S13, S17, S19; 42, 236), 5, 7a (3,6), 7c (S8, S13, S17, S19), 9a (M5, M12), 10 Abbott (1), Ambrus, Panel (3,7,8), Stijve (7)
028	dioxathion	2c, 2d, 4 (XII-6; S8, S13, S19), 7a (6), 7c (S8, S9, S19), 8e, 9a (M2, M5, M12), 10 Abbott (1), Stijve (7)
029	diphenyl	1f, 2d, 4 (XII-6; 256A), 7a (6), 10 Farrow, Kitada, Lord, Mestres (1), Player, Pyysalo
030	diphenylamine	2d, 2e, 4 (XII-6), 7a (6), 10 Allen (1), Luke
031	diquat	2e, 4 (37), 6d Calderbank (2), King
032	endosulfan	1b, 2a, 2d, 2f, 3, 4 (XII-5,6; S5, S8, S12, S19; 50), 5, 7a (5,6), 7c (S19), 5, 9a (M1, M12), 10 Abbott (2), Ambrus
033	endrin	1a, 1o, 2a, 2d, 2f, 3, 4 (XII-5,6; S5, S9, S10, S12, S19), 5, 7a (5,6), 7c (S9-10, S12, S19), 8a, 8b, 8c, 8d, 9a (M1, M12), 10 Abbott (2), Ambrus, Panel (4)
034	ethion	1a, 2a, 2c, 2d, 2f, 3, 4 (XII-3,5,6; S8, S9, S13, S17, S19), 7a (3,5,6), 7c (S8, S9, S13, S17, S19), 8e, 9a (M2, M5, M12), 10 Abbott (1), Ambrus, Stijve (7)

CCPR number	Compounds	References to Analytical Methods (see "References to Literature" to follow)
035	ethoxyquin	2d, 2e, 4 (XII-6; 500) Winell
036	fenchlorphos	1a, 2a, 2c, 2d, 2f, 3, 4 (XII-3,5,6; S8-10, S13, S17, S19), 7a (3,5,6), 7c (S8-10, S13, S17, S19), 8b, 8e, 9a (M2, M5), 10 Abbott (1), Ambrus, Panel (7,8), Stijve (7)
037	fenitrothion	2a, 2c, 2d, 2f, 3, 4 (XII-3,6; S5, S8, S13, S17, S19; 58), 6a, 8e, 9a (M2, M5), 10 Abbott (1), Ambrus, Bottomley, Desmarchelier, Panel (7,8), Stijve (7)
038	fensulfothion	2c, 2d, 2e, 3, 4 (XII-3,6; S8, S13, S16, S17, S19), 6a, 7a (3,6), 7c (S8, S13, S16, S17, S19), 9a (M2, M5), 10
039	fenthion	2c, 2d, 2e, 2f, 3, 4 (XII-3,6; S5, S8, S13, S16, S17, S19), 7a (3,6), 7c (S8, S13, S16, S17, S19), 8e, 9a (M2, M5), 10 Abbott (1), Ambrus, Hill
040	fentin	2e, 4 (S24; 55A, 55B), 6e Baker, PG (1)
041	folpet	2a, 2c, 2d, 3, 4 (XII-6; S8, S12, S19, S20; 91, 91A), 7a (6), 7b, 7c (S8, S12, S19, S20), 9a (M1, M12), 10 Ambrus, Baker, PB (1), Buettler, Gilvydis, Pomerantz
042	formothion	2d, 4 (XII-6; S5, S8, S19; 236), 6b, 7a (6), 7c (S8, S19), 9a (M2, M5, M12), 10 Abbott (1), Ambrus
043	heptachlor	1a, 1n, 1o, 2a, 2d, 2f, 3, 4 (XII-5,6; S 1-4, S8-10, S12, S19), 5, 6c, 6d, 7a (5,6), 7c (S8-10, S12, S19), 8a, 8b, 8c, 8d, 9a (M1, M12), 10 Abbott (2), Ambrus, Stijve (2,3), Veierov

CCPR number	Compounds	References to Analytical Methods (see "References to Literature" to follow)
044	hexachlorobenzene	1k, 1o, 2a, 2d, 3, 4 (XII-1,5,6; S9, S10, S12, S19), 5, 6c, 7a (1,5,6), 7c (S9, S10, S12, S19), 8a, 8b, 8c, 8d, 9a (M1, M12), 10 Ambrus, Panel (4), Stijve (2,3), Veierov, Zimmerli
045	hydrogen cyanide	2e, 4 (11), 9b Darr
046	hydrogen phosphide	2e, 4 (13), 9a (M8) Scudamore (2)
047	inorganic bromide	2e, 4 (S18; 149), 7c (S18), 9b Panel (2), Roughan, Stijve (1,4), VanWees
048	lindane	1a, 1o, 2a, 2d, 3, 4 (XII-5,6; S1-5, S8-10, S12, S19), 5, 7a (5,6), 7c (S8-10, S12, S19), 8a, 8b, 8c, 8d, 9a (M1, M12), 10 Abbott (2), Ambrus, Panel (4), Stijve (2,3), Veierov
049	malathion	1a, 2a, 2c, 2d, 2f, 3, 4 (XII-3,5,6; S5, S8, S10, S13, S17, S19; 72), 7a (3,5,6), 7c (S8, S10, S13, S17, S19), 8e, 9a (M2, M5, M12), 10 Abbott (1), Ambrus, Bottomley, Desmarchelier, Panel (1,3,7,8), Stijve (7)
050	mancozeb	see 105: dithiocarbamates
051	methidathion	2a, 2c, 2d, 2e, 3, 4 (XII-6; S5, S8, S13, S19; 232), 6b, 7a (6), 7c (S8, S13, S19), 9a (M2, M5, M12), 10 Ambrus
052	methyl bromide	9a (M8) Mestres (2), Panel (5)

CCPR number	Compounds	References to Analytical Methods (see "References to Literature" to follow)
053	mevinphos	2c, 2d, 2f, 3, 4 (XII-3,6; S5, S8, S13, S17, S19; 93), 7a (3,6), 7c (S8, S13, S17, S19), 9a (M2, M5, M12), 10 Abbott (1), Ambrus
054	monocrotophos	1p, 2c, 2d, 2e, 2f, 4 (XII-6; S19), 7c (S19), 9a (M2, M5), 10 Ambrus
055	omethoate	1p, 2c, 2d, 4 (XII-6; S13, S17, S19; 236), 5, 7a (6), 7c (S13, S17, S19), 9a (M2, M5), 10 Abbott (1), Panel (3)
056	ortho-phenylphenol	2d, 2e, 10 Farrow, Kitada, Lord, Mestres (1), Player, Pyysalo
057	paraquat	2e, 4 (134), 6d, 7b Calderbank (1), Khan, King, Lott
058	parathion	1a, 1c, 2a, 2c, 2d, 2f, 3, 4 (XII-3,4,5,6; S5, S8, S10, S13, S17, S19; 87A, 87B), 7a (3,4,5,6), 7c (S8, S10, S13, S17, S19), 8e, 9a (M2, M5, M12), 10 Abbott (1), Ambrus, Panel (3)
059	parathion-methyl	1a, 2a, 2c, 2d, 2f, 3, 4 (XII-3,5,6; S5, S8, S13, S17, S19; 88A, 88B), 7a (3,5,6), 7c (S8, S13, S17, S19), 8e, 9a (M2, M5, M12), 10 Abbott (1), Ambrus, Panel (3)
060	phosalone	2a, 2c, 2d, 2e, 3, 4 (XII-5,6; S8, S19), 5, 6a, 7a (5,6), 7c (S8, S19), 9a (M2, M5, M12), 10 Abbott (1), Ambrus, Stijve (7)
061	phosphamidon	2c, 2d, 2e, 3, 4 (XII-6; S5, S13, S19), 7a (6), 7c (S13, S19), 9a (M5, M12), 10 Abbott (1), Ambrus, Bottomley

CCPR number	Compounds	References to Analytical Methods (see "References to Literature" to follow)
62	piperonyl butoxide	1l, 2e, 4 (XII-6; S19, S22; 163), 7a (6), 7c (S19), 9b Krause (2)
063	pyrethrins	2a, 2d, 2e, 4 (XII-6; S19, S22), 6b, 7a (6), 7c (S19), 9b
064	quintozene	2a, 2d, 2f, 3, 4 (XII-4,5,6; S8, S9, S12, S19; 99), 7a (4,5,6), 7c (S8, S9, S12, S19), 9a (M1, M12), 10
065	thiabendazole	2d, 2e, 2h, 4 (XII-6; 256A, 256B), 8g, 9a (M3), 10 Farrow, Kitada, Mestres (1,3), Rajzman, Yamada
066	trichlorfon	2c, 2d, 2e, 2f, 3, 4 (XII-6; S5, S13, S19; 112), 5, 7a (6), 7c (S13, S19), 8e, 9a (M2, M5, M12) Abbott (1), Ambrus, Bottomley
067	cyhexatin	2e, 4 (S24), 6a, 9b Moellhoff (2)
068	azinphos-ethyl	2c, 2d, 4 (XII-3,5,6; S5, S8, S13, S17, S19; 62, 62A), 7a (3,5,6), 7c (S8, S13, S17, S19), 9a (M2, M5, M12), 10 Abbott (1), Ambrus
069	benomyl	see 072: carbendazim
070	bromopropylate	2a, 2d, 4 (XII-6; S19), 7a (6), 7c (S19), 9a (M12), 10 Stijve (6)
071	camphechlor	2a, 2d, 2e,4 (XII-5,6; S9, S19), 7a (5,6), 7c (S9, S19) Stijve (2)
072	carbendazim	2e, 2h, 4 (261, 378), 6a, 6d, 9a (M3), 10 Ambrus, Farrow, Mestres (3), VanHaver

CCPR number	Compounds	References to Analytical Methods (see "References to Literature" to follow)
073	demeton-S-methyl	2d, 2f, 4 (XII-6; S5, S13, S16, S19), 7a (6), 7c (S13, S16, S19), 9a (M2, M5), 10 Abbott (1), Ambrus, Hill, Wagner
074	disulfoton	2a, 2c, 2d, 2e, 2f, 3, 4 (XII-3,6; S5, S8, S13, S16, S17, S19), 7a (3,6), 7c (S8, S13, S16, S17, S19), 8e, 9a (M2, M5) Abbott (1), Ambrus, Panel (7)
075	propoxur	1e, 2d, 2g, 4 (XII-6; S19; S25; 216), 6a, 7a (6), 7c (S19), 9a (M2, M13), 10 Ambrus, Brauckhoff, Chaput, Lawrence(1)
076	thiometon	2d, 4 (XII-6;S13), 6b, 7a (6), 7c (S13), 9a (M2, M5, M10, M12) Abbott (1), Ambrus, Hill
077	thiophanate-methyl	2e, 2h, 4 (261), 5, 9a (M3), 10 Ambrus, Mestres (3), VanHaver
078	vamidothion	4 (XII-3,6; S17), 6a, 7a (3,6), 7c (S17), 9a (M2, M5, M10)
079	amitrole	2e, 4A Galoux, Løkke (1), v.d.Poll
080	chinomethionate	2d, 2e, 4 (XII-6; S19; 189), 7a (6), 7c (S19), 9b, 10 Ambrus, Francoeur, Krause (1), Tjan
081	chlorothalonil	2a, 2d, 2e, 3, 4 (XII-6; S19), 6b, 7a (6), 7c (S19), 9a (M1, M12), 10 Ambrus, Løkke (2)
082	dichlofluanid	2a, 2d, 4 (XII-6; S8, S12, S19; 203; 203A, 203 -(371)), 7a (6), 7c (S8, S12, S19), 9a (M1, M12), 10 Ambrus, Løkke (2), Brennecke (4)

CCPR number	Compounds	References to Analytical Methods (see "References to Literature" to follow)
083	dicloran	2d, 3, 4 (XII-6; S19), 7a (6), 7c (S19), 9a (M1), 10 Ambrus
084	dodine	1i, 2e Newsome (1)
085	fenamiphos	2c, 2d, 2e, 4 (XII-6; S8; S16; S19), 7a (6), 7c (S16, S19), 9a (M5, M12) Hill
086	pirimiphos-methyl	2a, 2c, 2d, 2e, 4 (XII-6; S8, S19; 476), 6b, 7a (6), 7c (S8, S19), 9a (M2, M5, M12), 10 Ambrus, Desmarchelier, Panel (7,8), Stijve (7)
087	dinocap	2a, 2d, 2e, 4 (XII-6; S19; 68), 7a (6), 7c (S19), 9a (M9), 9b Ambrus
088	leptophos	withdrawn
089	sec-butylamine	2e, 6b Day, Hunter, Scudamore (1)
090	chlorpyrifos-methyl	2c, 2d, 4 (XII-6; S8, S19), 7a (6), 7c (S19), 9a (M2, M5), 10 Ambrus, Bottomley, Desmarchelier, Panel (4,8), Stijve (7)
091	cyanofenphos	2d, 4 (XII-6; S8, S19), 7a (6), 7c (S19), 9a (M2, M5), 10
092	demeton	2c, 2d, 2e, 4 (XII-6; S5,S16), 7a (6), 7c (S16), 9a (M5) Abbott (1)
093	bioresmethrin	6c, 6d, 9a (M11) Baker, PG (2), Bottomley

CCPR number	Compounds	References to Analytical Methods (see "References to Literature" to follow)
094	methomyl	1q, 2d, 2e, 2g, 4 (299), 6a, 7b, 9a (M13) Ambrus, Chaput
095	acephate	1p, 2c, 2d, 2e, 4 (XII-6; S19; 358), 6a, 7a (6), 7b, 7c (S19), 9a (M5, M12), 10
096	carbofuran	1e, 1q, 2e, 2g, 3, 4 (XII-6; S25), 6a, 7a (6), 9a (M13), 10 Ambrus, Brauckhoff, Chaput, Lawrence(1), Moellhoff (1) Leppert (1,2)
097	cartap	Official Gazette
098	dialifos	2a, 2d, 2e, 4 (XII-6; S19; 281), 7a (6), 7c (S19), 9a (M2, M5, M12), 10
099	edifenphos	2d, 4 (XII-6; S19), 7a (6), 7c (S19)
100	methamidophos	1p, 2c, 2d, 3, 4 (XII-6; S19; 358, 365), 5, 6a, 7a (6), 7c (S19), 9a(M5), 10
101	pirimicarb	2d, 4 (XII-6; S19; 309), 5, 6a, 7b, 10
102	maleic hydrazide	1m, 4 (297) Lane, Newsome (3)
103	phosmet	2c, 2d, 4 (XII-6), 7a (6), 9a (M2, M5, M12), 10 Ambrus
104	daminozide	2e, 6b Allen (2), Newsome (5), Saxton, Wright, Conditt

CCPR number	Compounds	References to Analytical Methods (see "References to Literature" to follow)
105	dithiocarbamates	2e, 3, 4 (S15, S21), 7c (S21), 9b Newsome (2), Panel (6), Ott
106	ethephon	2e, 9b Cochrane
107	ethiofencarb	2d, 2g, 4 (S25; 393), 9a (M13), 10
108	ethylene thiourea	1j, 4 (389), 7b, 9b Panel (9), Hirvi, Otto, Rosenberg
109	fenbutatin oxide	2e, 4 (S24), 6d Sano
110	imazalil	2d, 2e, 4 (XII-6; S19)
111	iprodione	2c, 2d, 2e, 4 (XII-6; S8, S19; 419), 6e, 7a (6), 7c (S8, S19), 9a (M1, M12), 10
112	phorate	2a, 2c, 2d, 2e, 4 (XII-3,6; S8, S13, S16, S17, S19), 7a (3,6), 7c (S8, S13, S16, S17, S19), 9a (M2, M5) Abbott (1), Ambrus, Hill
113	propargite	2a, 2d, 2e, 3, 4 (XII-6), 6a, 7a (6), 9a (M1) Ambrus
114	guazatine	Kobayashi
115	tecnazene	2a, 2d, 2e, 3, 4 (XII-6; S8, S12, S19; 108), 7a (6), 7c (S8, S12, S19), 9a (M1), 10

CCPR number	Compounds	References to Analytical Methods (see "References to Literature" to follow)
116	triforine	2e, 4 (338), 6d, 9b Bourke, Newsome (4)
117	aldicarb	1q, 2e, 2g, 4 (XII-6; 250), 6a, 7a (6), 9a (M10, M13), 10 Ambrus, Chaput
118	cypermethrin	2a, 2d, 4 (XII-6; S19, S23), 6g, 7a (6), 7c (S19), 9a (M11), 10 Ambrus, Baker, PG (2), Bottomley
119	fenvalerate	2a, 2d, 2e, 4 (XII-6; S19, S23), 6g, 7a (6), 7c (S19), 9a (M11), 10 Ambrus, Baker, PG (2), Bottomley
120	permethrin	2a, 2d, 2e, 4 (XII-6; S19, S23), 6g, 7a (6), 7c (S19), 9a (M11), 10 Ambrus, Baker, PG (2), Bottomley
121	2,4,5-T	2b, 4 (XII-6; 105), 6c, 7a (6), 9a (M6) Ebing, Løkke (3), Specht (1)
122	amitraz	2e, 4 (XII-6), 7a (6), 9b
123	etrimfos	2a, 2c, 2d, 4 (XII-6; S8, S19), 7a (6), 7c (S19), 6e, 9a (M2, M5) Ambrus, Bottomley, Panel (8)
124	mecarbam	2c, 2d, 4 (XII-6; S19), 6b, 7a (6), 7c (S19), 9a (M2),10 Abbott (1)
125	methacrifos	4 (XII-6), 7a (6) Ambrus, Desmarchelier, Panel (7,8)

CCPR number	Compounds	References to Analytical Methods (see "References to Literature" to follow)
126	oxamyl	1q, 2e, 2g, 4 (XII-6; 441), 5, 7a (6), 9a (M13), 10 Ambrus
127	phenothrin	4 (XII-6), 7a (6), 9 Baker, PG (2), Bottomley
128	phenthoate	2a, 2c, 2d, 4 (XII-6; S19), 6b, 7a (6), 7c (S19), 9a (M11), 10 Ambrus
129	azocyclotin	4 (S24) Moellhoff (2)
130	diflubenzuron	2e, 6d, 6f, 9a (M4) Austin
131	isofenphos	2a, 2c, 2d, 2e, 4 (XII-6; S8), 7a (6), 9a (M5, M12), 10
132	methiocarb	1q, 2d, 2g, 4 (79, 79A), 9a (M2, M13), 10 Chaput
133	triadimefon	2d, 2e, 4 (XII-6; S8, S19; 425-(605)), 7a (6), 7c (S8, S19), 10 Ambrus, Brennecke (2), Ragab
134	aminocarb	2d, 10 Brauckhoff
135	deltamethrin	2a, 2d, 4 (XII-6; S19, S23), 6g, 7a (6), 7c (S19), 9a (M11) Ambrus, Baker, PG (2), Bottomley
136	procymidone	2a, 2d, 4 (XII-6; S8, S19), 7a (6), 7c (S8, S19), 10

CCPR number	Compounds	References to Analytical Methods (see "References to Literature" to follow)
137	bendiocarb	2d, 2g, 6d, 4 (XII-6), 7a (6), 9a (M2, M13) Ambrus
138	metalaxyl	2c, 2d, 2e, 4 (XII-6; S8, S19; 517), 7a (6), 7b, 7c (S19), 9a (M4), 10 Ambrus
139	butocarboxim	2g, 9a (M13) Aharonson, Brauckhoff, Li, Muszkat
140	nitrofen	1a, 2a, 2d, 2e, 4 (XII-6; S19; 340), 6d, 7a (6), 7b, 7c (S19) Adler, Ambrus, Yu
141	phoxim	2d, 4 (XII-6; S19; 307), 7a (6), 7c (S19), 9a (M2, M12) Ambrus
142	prochloraz	2d Maclaine Pont, Somerville
143	triazophos	2c, 2d, 4 (XII-4,6; S8, S19; 401), 6d, 7a (6), 7c (S19), 9a (M2, M5, M12), 10 Ambrus
144	bitertanol	2d, 4 (XII-6; S19; 613; 613A), 7a (6), 7c (S19), 9a (M12) Brennecke (1,3)
145	carbosulfan	2d, 4 (658 - (344)) Leppert (1,2)
146	cyhalothrin	2d, 6g
147	methoprene	2e, 6d

CCPR number	Compounds	References to Analytical Methods (see "References to Literature" to follow)
148	propamocarb	Gentile
149	ethoprofos	2c, 2d, 2e, 4 (XII-6; S8, S19), 7a (6), 7b, 7c (S19), 9a (M2, M5) Ambrus
150	propylene thiourea	Lembo, Nitz
151	dimethipin	2e
152	flucythrinate	2d, 2e
153	pyrazophos	2d, 4 (XII-4,6; S8, S19; 328), 6d, 7a (6), 7b, 7c (S19), 9a (M2, M5, M12), 10
154	thiodicarb	2g
155	benalaxyl	4(S19) not published yet
156	clofentezine	
157	cyfluthrin	2d, 4 (S23), 9a (M11)
158	glyphosate	2e, 4 (405) (10th ed.), 9b Dubelman, Tuinstra, Wigfield
159	vinclozolin	2a, 2d, 4 (XII-6; S8, S19; 412), 9a (M1, M12)
160	propiconazole	2d, 4 (S19; 624)

CCPR number	Compounds	References to Analytical Methods (see "References to Literature" to follow)
161	paclobutrazol	2d Reed
162	tolylfluanid	4 (XII-6; S 8; S19: 371; 203- (371)), 7c (S8, S19), 9a (M1,M12), 2d Brennecke (4) Specht (2), Anderson
163	anilazine	4 (XII-6; S19: 186), 7c (S19), 2d, 2e Lawrence(2), Brennecke(5)
164	demeton-S-methyl sulphone	4(XII-6, S16, S19), 7c (S16), 9a (M5), 2d, 2e Andersson, Thornton, Wagner
165	flusilazole	2d, 4(S19)(only parent compound)
166	oxydemeton-methyl	4(XII-6, S16, S19), 7c (S16), 9a (M5), 2c, 2d, 2e Thornton, Wagner
167	terbufos	4 (S8; S19), 9a(M5) (only parent compound), 2c, 2d, 2e, Westcott
168	triadimenol	4 (XII-6, S19, 425 - (605)) 7a (6), 7c (S19), 9a (M12), 2d, Allmendinger, Andersson, Brennecke (2), Ragab, Mendes
169	cyromazine	2e Cabras, Bardalaye
170	hexaconazole	
171	profenofos	2c, 2d, 2e Andersson

REFERENCES TO LITERATURE

GENERAL ARTICLES

The following papers or books deal with general problems in pesticide residue analysis and are not used as references for the preceeding "List of Methods of Analysis".

Ambrus, A. & Thier, H.-P., Application of multi-residue procedures in pesticides residues analysis, Pure Appl. Chem., 58, 1035-1062 (1986)

Beck, H., Untersuchungsmethoden zur Bestimmung der Rückstände von Chlorkohlenwasserstoff-Pestiziden in oder auf Lebensmitteln, Bundesgesundheitsblatt, 17, 269-274 (1974)

Becker, G. et al., Dünnschichtchromatographie in der Rückstandsanalytik von Pflanzenschutzmitteln und deren Metaboliten, Verlag Chemie VCH, Weinheim, FRG (1987)

Burke, J.A., The interlaboratory study in pesticide residue analyses, in: Advances in Pesticide Science, H. Geissbuehler (edit.), Pergamon Press, Oxford, UK, 633-642 (1979)

Carl, M., Internal laboratory quality control in the routine determination of chlorinated pesticides,in: Advances in Pesticide Science, H. Geissbuehler (edit.), Pergamon Press, Oxford, UK, 660-663 (1979)

Cochrane, W.P., Chemical derivatization in pesticide analysis, Plenum Press, New York, N.Y., USA, (1981)

Egli, H., Storage stability of pesticide residues, J. Agr. Fd. Chem., 30, 861-866 (1982)

Frehse, H. & Timme, G., Quantitative residue analytical reliability: beatitude through the application of latitude, Res. Revs., 73, 27-47 (1980)

Gunther, F.A., Interpreting pesticide residue data at the analytical level, Res. Revs., 76, 155-171 (1980)

Horwitz, W., The role of the analyst in analytical chemistry, FDA Bylines, 4, 169-178 (1979)

Horwitz, W., The inevitability of variability in pesticide residue analysis, in: Advances in Pesticide Science, H. Geissbuehler (edit.), Pergamon Press, Oxford, UK, 649-655 (1979)

Horwitz, W. et al., Quality assurance in the analysis of foods for trace constituents, JAOAC, 63, 1344-1354 (1980)

Horwitz, W., Evaluation of analytical methods used for regulation of foods and drugs, Anal. Chem., 54, 67A-76A (1982)

ISO Document ISO 5725, 2nd edit. (1986), Precision of test methods: determination of repeatability and reproducibility

IUPAC Reports on Pesticides (13), Development and evaluation of simplified approaches to residues analysis, Pure Appl. Chem., 53, 1039-1049 (1981)

Moye, H.A. (edit.), Analysis of pesticide residues, Vol. 58 of: Chemical Analysis, John Wiley and Sons, New York, N.Y., USA (1981)

Pesticide Residue Analysis, Health Aspects of Chemical Safety, Interim Document 14, WHO, Regional Office for Europe, Copenhagen, Denmark (1984)

Safe, S. & Hutzinger, O., Mass spectrometry of pesticides and pollutants, CFC Press Inc., Boca Raton, Florida, USA (1979)

Smart, N., Samples used for interlaboratory studies of methods for pesticide residues analysis in foodstuffs, Res. Revs., 96, 1-12 (1985)

Steiner, E.H., Planning and analysis of results of collaborative tests, in: Statistical Manual of the AOAC, Washington, D.C., USA (1974)

The Agrochemical Handbook, The Royal Society of Chemistry, The University, Nottingham, UK, (1983)

Thier, H.-P. & Frehse, H., Rückstandsanalytik von Pflanzenschutzmitteln, Georg Thieme Verlag, Stuttgart - New York (1986)

Youden, W.J., Statistical techniques for collaborative tests, in: Statistical Manual of the AOAC, Washington, D.C., USA (1974)

Zweig, G. (edit.), Analytical methods for pesticides, plant growth regulators and food additives, Academic Press, New York - San Francisco - London, Vol. XIV and XV (1986)

MANUALS

Included in this listing are those international and national laboratory analysis publications and manuals selected as references for the preceeding "List of Methods of Analysis".

(1)   Official Methods of Analysis of the Association of Official Analytical Chemists, 15th edition (1990)

     a.   970.52
     b.   976.23
     c.   974.22
     d.   977.18
     e.   975.40
     f.   968.25
     g.   957.14
     h.   964.18
     i.   964.19
     j.   978.16
     k.   977.19
     l.   960.43
     m.   963.24
     n.   983.21
     o.   984.21
     p.   985.22
     q.   985.23

(2)   Pesticide Analytical Manual, Food and Drug Administration, Washington, D.C., USA

     a.   Vol. I, Table 201-A and sections, 211.1, 212.1, 231.1, 232.1 and 252
     b.   Vol. I, Table 201-D and section 221.1
     c.   Vol. I, Table 201-H and section 232.3
     d.   Vol. I, Table 201-I and section 232.4
     e.   Vol. II, Method under compound name (when in this reference several methods have been given, they are generally listed in order of preference)
     f.   Vol. I, Table 651-A and sections 650 and 651
     g.   Vol. I, Table 242.2-1 and section 242.2
     h.   Vol. I, Section 242.3

(3)   Manual on Analytical Methods for Pesticide Residues in Foods, Health Protection Branch, Health and Welfare Canada, Ottawa, Ont., Canada (1985) (available in English and French)

(4)   Methodensammlung zur Rückstandsanalytik von Pflanzenschutzmitteln, 1.- 11. Lieferung, Verlag Chemie GmbH, Weinheim/Bergstrasse, FRG (1991) (the numbers in parentheses refer to the numbers of the methods in this manual; methods preceeded by "S" are multi-residue methods; the manual is also available in English, see ref. 7, below)

(5) Laboratory Manual for Pesticide Residues Analysis in Agricultural Products, compiled by R.B. Maybury, Pesticide Laboratory, Food Production and Inspection Branch, Agriculture Canada, Ottawa, Ont., Canada (1984) (available in English and French)

(6) Zweig, G. (edit.), Analytical Methods for Pesticides, Plant Growth Regulators and Food Additives, Academic Press, New York - San Francisco - London

   a. Vol. VII (1974)
   b. Vol. VIII (1976)
   c. Vol. IX (1977)
   d. Vol. X   (1978)
   e. Vol. XI (1980)
   f. Vol. XII (1982)
   g. Vol. XIII (1984)

(7) Manual of Pesticide Residue Analysis, Deutsche Forschungsgemeinschaft, VCH Verlagsgesell-schaft, Weinheim, FRG (1987) (English translation of ref. 4, above)

   a. Vol. I, Section Clean-up Methods (the numbers in parentheses refer to the numbers of the clean-up methods in this volume)
   b. Vol. I, Section Individual Pesticide Residue Analytical Methods
   c. Vol. I, Section Multiple Pesticide Residue Analytical Methods (the numbers in parentheses refer to the numbers of the multi-residue methods in this volume)

(8) Chemistry Laboratory Guidebook, United States Department of Agriculture, Food Safety and Inspection Service, Science Program, Washington, D.C., USA

   a. Section 5.001
   b. Section 5.002
   c. Section 5.003
   d. Section 5.004
   e. Section 5.006
   f. Section 5.005
   g. Section 5.050

(9) Analytical Methods for Residues of Pesticides in Foodstuffs, P.A. Greve (edit.), 5th edition, Government Publishing Office, The Hague, Netherlands (1988)

   a. Part I: Multi-residue Methods (the numbers in parentheses refer to numbers of the multi-residue methods in this volume)
   b. Part II: Special Methods (methods given under compound name)

(10) Materials and Methods Used for Pesticide Residues Monitoring in Sweden, Vår Föda, 38, Suppl.2, 79-136 (1986)

INDIVIDUAL PAPERS

These are selected references from the scientific literature, which refer to analysis of the residues of specified pesticide chemicals. The references, by name of the (first) author are used in the preceeding "List of Methods of Analysis". As a further cross-reference, the CCPR numbers of the pesticide(s) to which a given paper applies, are listed in italics after each reference. The format is therefore: Name of author and paper reference; title of paper; and CCPR numbers of applicable pesticides.

Abbott (1), D.C. et al., Pest. Sci., 1, 10-13 (1970)
Pesticide residues in the total diet in England and Wales, 1966-1967; Part III: Organophosphorus pesticide residues in the total diet
2, 4, 5, 11, 14, 22, 25, 27, 28, 34, 36, 37, 39, 42, 49, 53, 55, 58, 59, 60, 66, 68, 73, 74, 76, 92, 112, 124

Abbott (2), D.C. et al., J. Chromatog., 16, 481-487 (1964)
Some observations on the thin-layer chromatography of organochlorine pesticides
1, 21, 32, 33, 43, 48

Adachi, K. et al., JAOAC, 67, 798-800 (1984)
Simple analytical method for organophosphorus pesticide determination in unpolished rice, using removal of fats by zinc acetate
22, 27, 37, 49, 58, 128

Adler, I.L. & Wargo Jr, J.P., JAOAC, 58, 551-553 (1975)
Determination of residues from the herbicide 2,4-dichloro-1-(4-nitrophenoxy)-benzene in rice and wheat by electron-capture gas-liquid chromatography
140

Aharonson, N. & Muszkat, L., Z. Lebensm. Unters. Forsch., 180, 96-100 (1985)
Direct gas chromatographic determination of the two isomeric insecticides, aldicarb and butocarboxime and their toxic metabolites: application to residue analysis in crops and leaves
117, 139

Allen (1), J.G. & Hall, K.J., J. Agr. Fd. Chem., 28, 255-258 (1980)
Methods for the determination of diphenylamine residues in apples
30

Allen (2), J.G., Pest. Sci., 11, 347-350 (1980)
Daminozide residues in sweet cherries, and their determination by colorimetric and gas-liquid chromatographic methods
104

Allmendinger, H. Pflanzensch. Nachr. Bayer, 44, 5-66 (1991)
A method for determining residues of the fungicides Folicur[R] and Bayfidan[R] in plant material and soil by gas chromatography.
168

Ambrus, A. et al., JAOAC, 64, 733-768 (1981)
General method for determination of pesticide residues in samples of plant origin, soil, and water
1, 2, 4, 6, 7, 8, 14, 17, 21, 22, 25, 27, 32, 33, 37, 39, 41, 42, 43, 48, 49, 51, 53, 54, 58, 59, 60,
61, 66, 72, 73, 74, 75, 76, 77, 80, 81, 82, 83, 86, 87, 94, 96, 101, 103, 112, 113, 117, 118, 119,
120, 123, 128, 133, 135, 137, 140, 141, 143, 159

Andersson, A and Ohling, B. Var Föda 38, 79-109 (1986)
A capillary gas chromatographic multiresidue method for the determination of fruit and vegetables.
162, 164, 168, 171

Austin, D.J. & Hall, K.J., Pest. Sci., 12, 495-502 (1981)
A method of analysis for the determination of binapacryl, bupirimate and diflubenzuron on apple foliage
and fruit, and its application to persistence studies
3, 130

Baker, P.B. (1) & Flaherty, B., Analyst, 97, 713-718 (1972)
Fungicide residues; Part II: The simultaneous determination of residues of folpet, captan and captafol
in selected fruits by gas chromatography
6, 7, 41

Baker, P.B. (2) & Hoodless, R.A., Analyst, 98, 172-175 (1973)
Fungicide residues; Part III: The determination of binapacryl in selected fruits by gas chromatography
3

Baker, P.G. (1) et al., Analyst, 105, 282-285 (1980)
Fungicide residues; Part VII: Determination of residues of fentin in vegetables and cocoa products by
spectrofluorimetry
40

Baker, P.G. (2) & Bottomley, P., Analyst, 107, 206-212 (1982)
Determination of residues of synthetic pyrethroids in fruit and vegetables by gas-liquid and
high-performance liquid chromatography
93, 118, 119, 120, 127, 135

Bardalaye, C, Wheeler, W.B. & Meister C.W. JAOAC, 70, 455-457 (1987)
Gas chromatographic determination of cyromazine and its degradation product melamine in chinese
cabbage.
169

Bottomley, P. & Baker, P.G., Analyst, 109, 85-90 (1984)
Multi-residue determination of organochlorine, organophosphorus and synthetic pyrethroid pesticides
in grain by gas-liquid and high-performance liquid chromatography
1, 4, 21, 22, 25, 37, 44, 49, 61, 66, 86, 90, 93, 118, 119, 120, 123, 127, 135

Bourke, J.B. et al., J. Agr. Fd. Chem., 25, 36-39 (1977)
Residues and disappearance of triforine from various crops
116

Brauckhoff, S. & Thier, H.-P., Z. Lebensm. Unters. Forsch., 184, 91-95 (1987)
Analysenmethode für Rückstände von Methylcarbamat-Insecticiden in pflanzlichen Lebensmitteln
8, 75, 94, 96, 101, 107, 117, 132, 134, 137, 139

Brennecke (1), R., Pflanzensch. Nachr. Bayer, (Engl. edit.)38, 33-54 (1985)
Method for gas-chromatographic determination of residues of Baycor[R] fungicide in plant material, soil
and water
144

Brennecke (2), R., Pflanzensch. Nachr. Bayer, 37, 68-93 (1984)
Method for gas-chromatographic determination of residues of Bayleton[R] and Bayfidan[R] fungicides in
plant material, soil and water (German edit.: 37, 66-91 (1984))
133, 168

Brennecke (3), R., Pflanzensch. Nachr. Bayer , 41, 113-131 (1988)
Method for the determination of residues of the fungicide Baycor[R] in plant material and beverages by
high pressure liquid chromatography with fluorescence detection (German edit.: 41, 113-135 (1988))
144

Brennecke (4), R., Pflanzensch. Nachr. Bayer, 41, 137-174 (1988).
A method for the determination of residues of the fungicides Euparen[R] and Euparen M[R] in plant material
and beverages by gas chromatography (German edit. 41, 136-172 (1988)).
82, 162

Brennecke (5), R., Pflanzensch. Nachr. Bayer, 38, 11-32 (1985)
Method for gas-chromatographic determination of Dyrene[R] residues in plant material, soil and water
(German edit.: 38, 11-32 (1985)).
163

Buettler, B. & Hoermann, W.D., J. Agr. Fd. Chem., 29, 257-260 (1981)
High-pressure liquid chromatographic determination of captan, captafol, and folpet residues in plant
material
6, 7, 41

Cabras, P., Meloni, M., & Spaneddal, J. Chromatogr., 505, 413-416, 1990
High-performance liquid chromatographic separation of cyromazine and its metabolite melamine.
169

Calderbank (1), A. & Yuen, S.H., Analyst, 90, 99-106 (1965)
An ion-exchange method for determining paraquat residues in food crops
57

Calderbank (2), A. & Yuen, S.H., Analyst, 91, 625-629 (1966)
An improved method for determining residues of diquat
31

Chaput, D. JAOAC, 71, 542-546, (1988)
Simplified multiresidue method for liquid chromatographic determination of N-methylcarbamate
insecticides in fruits and vegetables.
8, 75, 94, 96, 117, 132

Cochrane, W.P. et al., JAOAC, 59, 617-621 (1976)
Gas-liquid chromatographic analysis of ethephon and fenoprop residues in apples and their decline
before and after harvest
106

Conditt, M et al, JAOAC, 71, 735-739 (1988).
Gas chromatography/mass spectrometric determination of daminozide in high protein food products.
104

Daft, J.L., JAOAC, 66, 228 (1983)
Gas chromatographic determination of fumigant residues in stored grains, using isooctane partitioning
and dual column packings
10, 23, 24

Day, E.W. et al., JAOAC, 51, 39-44 (1968)
Determination of sec-butylamine residues in fruit
89

Desmarchelier, J. et al., Pest. Sci., 8, 473-483 (1977)
A collaborative study of residues on wheat of chlorpyrifos-methyl, fenitrothion, malathion, methacrifos
and pirimiphos-methyl
37, 49, 86, 90, 125

Dubelman, S. Analytical Methods for Pesticides and Plant Growth Regulators, Sherma, J, (editor)
Acad. Press Inc., San Diego, California, Vol XVI, 69-82 (1988).
Glyphosate
158

Ebing, W. et al., Lebensm. gerichtl. Chem., 39, 126-130 (1985)
Zur Rückstandsanalytik von Phenoxyalkancarbonsäure-Herbiziden in Getreidekörnern
20, 121

Farrow, J.E. et al., Analyst, 102, 752-758 (1977)
Fungicide residues; Part VI: Determination of residues of post-harvest fungicides on citrus fruit by
high-performance liquid chromatography
29, 56, 65, 72

Francoeur, Y. & Mallet, V., JAOAC, 59, 172-173 (1976)
Determination of quinomethionate (6-methylquinoline-2,3-diyldithiocarbamate) residues in crops by in situ fluorimetry
80

Galoux, M. et al., JAOAC, 65, 24-27 (1982)
Colorimetric determination of 3-amino-1,2,4-triazole in grain or meal
79

Gentile, I.A. & Passera, E., J. Chromatog., 236, 254-257 (1982)
Separation and detection of propamocarb by thin-layer chromatography
148

Gilvydis, D.M. & Walters, S.M., JAOAC, 67, 909-912 (1984)
Determination of captan, folpet, and captafol in fruits and vegetables, using two multiresidue methods
6, 7, 41

Heikes, D.L., JAOAC, 68, 431-436 (1985)
Purge and trap method for determination of ethylene dibromide in table-ready foods
23

Hill, A.R.C. et al., Analyst, 109, 483-487 (1984)
Organophosphorus sulphides, sulphoxides and sulphones; Part I: Determination of residues in fruit and vegetables by gas-liquid chromatography
38, 39, 73, 74, 76, 85, 112

Hirvi, T. et al., J. Agr. Fd. Chem., 27, 194-195 (1979)
A glass capillary gas-liquid chromatography method for determining ethylenethiourea without derivatization
108

Hunter, K. & Lindsay, D., Pest. Sci., 12, 319-324 (1981)
High-pressure liquid chromatographic determination of sec-butylamine residues in potatoes
89

Khan, S.U., Bull. Envir. Cont. Tox., 14, 745-749 (1975)
Determination of paraquat residues in food crops by gas chromatography
57

King, R.R., J. Agr. Fd. Chem., 26, 1460-1463 (1978)
Gas chromatographic determination of diquat residues in potato tubers
31

Kitada, Y. et al., JAOAC, <u>65</u>, 1302-1304 (1982)
Simultaneous liquid chromatographic determination of thiabendazole, o-phenylphenol, and diphenyl residues in citrus fruits, without prior cleanup
29, 56, 65

Kobayashi, H. et al., J. Pest. Sci., <u>2</u>, 427-430 (1977)
Gas chromatographic determination of guanidino fungicide, guazatine, in rice grain
114

Krause (1), R.T. & August, E.M., JAOAC, <u>66</u>, 1018-1022 (1983)
Applicability of a multiresidue method and high performance liquid chromatography for the determination of chinomethionate in apples and oranges
80

Krause (2), R.T. & August, E.M., JAOAC, <u>66</u>, 234-240 (1983)
Applicability of a carbamate insecticide multiresidue method for determining additional types of pesticides in fruits and vegetables
62

Lane (1), J.R., JAOAC, <u>46</u>, 261-268 (1963)
Collaborative study of maleic hydrazide residue analysis
102

Lane (2), J.R., JAOAC, <u>48</u>, 744-748 (1965)
Collaborative study of maleic hydrazide residue analysis
102

Lawrence(1), J.F., J. Agr. Fd. Chem., <u>25</u>, 211-212 (1977)
Direct analysis of some carbamate pesticides in foods by high-pressure liquid chromatography
8, 75, 96

Lawrence (2), J.F. & Panopio, L.G. JAOAC, <u>63</u>, 1300-1303 (1980)
Comparison of gas and liquid chromatography for determination of anilazine in potatoes and tomatoes.
163

Lembo, S. et al., J. Chromatog., <u>267</u>, 427-430 (1983)
Gas-liquid chromatographic method for determining propylenethiourea in rat tissues and fluids
150

Leppert (1), B.C. et al., J. Agr. Fd. Chem., <u>31</u>, 220-223 (1983)
Determination of carbosulfan and carbofuran residues in plants, soil, and water by gas chromatography
145

Leppert (2), B.C. et al., J. Agr. Fd. Chem., <u>32</u>, 1441 (1984)
Comment on recovery of carbosulfan residues from acidic crops
145

Li Yu-Chang et al., Fres. Z. Anal. Chem., 316, 290-292 (1983)
Methode zur Bestimmung von Rückständen an Butocarboxim in Pflanzen und Boden mit Hilfe der HPLC
139

Løkke (1), H., J. Chromatog., 200, 234-237 (1980)
Determination of amitrole by ion-pair high-performance liquid chromatography
79

Løkke (2), H., J. Chromatog., 179, 259-270 (1979)
Investigation on loss of chlorothalonil, dichlofluanid, tolylfluanid and vinclozolin by column chromatography clean-up on silver-loaded alumina in a gas chromatographic multiresidue procedure
81, 82

Løkke (3), H. & Odgaard, P., Pest. Sci., 12, 375-384 (1981)
Residues in blackcurrants, fodder peas, spinach and potatoes treated with sublethal doses of 2,4,5-T to simulate wind drift damage
121

Lord, E. et al., J. Assoc. Publ. Anal., 16, 25-32 (1978)
The determination of biphenyl and 2-hydroxybiphenyl in citrus fruit
29, 56

Lott, P.F. et al., J. Chromat. Sci., 16, 390-395 (1978)
The determination of paraquat
57

Love, J.L. & Patterson, J.E., JAOAC, 61, 627-628 (1978)
Atomic absorption spectrometric determination of cyhexatin
67

Lubkowitz, J.A. et al., J. Agr. Fd. Chem., 21, 143-144 (1973)
Residue studies of O,S-dimethyl phosphoroamidothioate on tomatoes
100

Luke, B.G. & Cossens, S.A., Bull. Envir. Cont. Tox., 24, 746-751 (1980)
Determination of diphenylamine residues in apples
30

Maclaine Pont, M.A. et al., Meded. Fac. Landbouww. Rijksuniv. Gent, 45, 835-840 (1980)
The residue analysis of prochloraz in combination with dicloran
142

Mendes, M.C.S. J. Agric. Fd. Chem., 38 174-178 (1990)
Evaluation and confirmation of acetylation gas liquid chromatographic method for the determination of triadimenol in foods.
168

Mestres (1), R. et al., Trav. Soc. Pharm. Montpellier, 35, 87-100 (1975)
Méthode rapide de controle et de dosage des résidus d'ortho-phényl phénol et de biphényle dans les agrumes
29, 56, 72, 77

Mestres (2), R. et al., Ann. Fals. Exp. Chim., 73, 407-420 (1980)
Méthode de recherche et de dosage des résidus de pesticides dans les produits céréaliers; 2°: Fumigants
9, 10, 23, 24, 52

Mestres (3), R. et al., Proc. Int. Soc. Citricult., 3, 1103-1106 (1977)
Thiophanate-methyl postharvest residues in oranges
65, 72, 77

Moellhoff (1), E., Pflanzensch. Nachr. Bayer (Engl. edit.), 28, 370-381 (1975)
Method for gas-chromatographic determination of Curaterr[R] residues in plants and soil samples with consideration to metabolites
96

Moellhoff (2), E., Pflanzensch. Nachr. Bayer (Engl. edit.), 30, 249-263 (1977)
Method for gas-chromatographic determination of Peropal[R] acaricide and its metabolites in plants, soil, water and laboratory animal chow
67, 129

Muszkat, L. & Aharonson, N., J. Chromat. Sci., 21, 411-414 (1983)
GC/CI/MS analysis of aldicarb, butocarboxime, and their metabolites
117, 139

Newsome (1), W.H., J. Agr. Fd. Chem., 24, 997-999 (1976)
A gas-liquid chromatographic method for the determination of dodine residues in foods
84

Newsome (2), W.H., J. Agr. Fd. Chem., 22, 887-889 (1974)
A method for determining ethylenebis(dithiocarbamate) residues on food crops as bis(trifluoroacetamido) ethane
105

Newsome (3), W.H., J. Agr. Fd. Chem., 28, 270-272 (1980)
A method for the determination of maleic hydrazide and its b-D-glucoside in foods by high-pressure anion-exchange liquid chromatography
102

Newsome (4), W.H., J. Agr. Fd. Chem., 30, 778-779 (1982)
Determination of triforine in fruit crops as N,N'-bis(pentafluorobenzoyl)piperazine
116

Newsome (5), W.H., J. Agr. Fd. Chem., 28, 319-321 (1980)
Determination of daminozide residues on foods and its degradation to 1,1-dimethylhydrazine by cooking
104

Nitz, S. et al., J. Agr. Fd. Chem., 30, 593-596 (1982)
A capillary gas-liquid chromatographic method for determination of ethylenethiourea and propylenethiourea in hops, beer and grapes
108, 150

Official Gazette, no. 4 of the Notification issued on March 20, 1979, by the Japan Environment Agency
Residue analysis of cartap hydrochloride
97

Ott, D.E. & Gunther, F.A., JAOAC, 65, 909-912 (1982)
Field screening method for above-tolerance residues of dithiocarbamate fungicides
105

Otto, S. et al., J. Envir. Sci. Health, Part B, 12, 179-191 (1977)
A new gas chromatographic determination of ethylene thiourea residues without derivatization
108

Panel (1) of the Committee for Analytical Methods for Residues of Pesticides and Veterinary Products
in Foodstuffs of the Ministry of Agriculture, Fisheries and Food, Analyst, 98, 19-24 (1973)
The determination of malathion and dichlorvos residues in grain
25, 49

Panel (2) of the Committee for Analytical Methods for Residues of Pesticides and Veterinary Products
in Foodstuffs of the Ministry of Agriculture, Fisheries and Food, Analyst, 101, 386-390 (1976)
Determination of residues of inorganic bromide in grain
4

Panel (3) of the Committee for Analytical Methods for Residues of Pesticides and Veterinary Products
in Foodstuffs of the Ministry of Agriculture, Fisheries and Food, Analyst, 102, 858-868 (1977)
Determination of residues of organophosphorus pesticides in fruits and vegetables
2, 27, 49, 55, 58

Panel (4) of the Committee for Analytical Methods for Residues of Pesticides and Veterinary Products
in Foodstuffs of the Ministry of Agriculture, Fisheries and Food, Analyst, 104, 425-433 (1979)
Determination of organochlorine pesticides in animal fats and eggs
1, 21, 33, 44, 48

Panel (5) of the Committee for Analytical Methods for Residues of Pesticides and Veterinary Products
in Foodstuffs of the Ministry of Agriculture, Fisheries and Food, Analyst, 99, 570-576 (1974)
The determination of residues of volatile fumigants in grain
10, 23

Panel (6) of the Committee for Analytical Methods for Residues of Pesticides and Veterinary Products
in Foodstuffs of the Ministry of Agriculture, Fisheries and Food, Analyst, 106, 782-787 (1981)
Determination of residues of dithiocarbamate pesticides in foodstuffs by a headspace method
105

Panel (7) of the Committee for Analytical Methods for Residues of Pesticides and Veterinary Products
in Foodstuffs of the Ministry of Agriculture, Fisheries and Food, Analyst, 105, 515-517 (1980)
Determination of a range of organophosphorus pesticide residues in grain
4, 22, 27, 37, 49, 74, 86, 112

Panel (8) of the Committee for Analytical Methods for Residues of Pesticides and Veterinary Products
in Foodstuffs of the Ministry of Agriculture, Fisheries and Food, Analyst, 110, 765-768 (1985)
Determination of a range of organophosphorus pesticide residues in grain
4, 27, 37, 49, 86, 90, 123, 125

Panel (9) of the Committee for Analytical Methods for Residues of Pesticides and Veterinary Products
in Foodstuffs of the Ministry of Agriculture, Fisheries and Food, Analyst, 112, 1559-1563 (1987)
Determination of ethylenethiourea in canned fruits and vegetables
108

Player, R.B. & Wood, R., J. Assoc. Publ. Anal., 18, 109-117 (1980)
Methods of analysis - collaborative studies; Part III: Determination of biphenyl and 2-hydroxy- biphenyl
in citrus fruit
29, 56

van der Poll, J.M., Vink, M. and Quirijns, J.K. Chromatographia, 30, 155-158, 1990.
Determination of amitrole in plant tissues and sandy soils by capillary gas chromatography with alkali
flame ionization detection.
79

Pomerantz, I.H. & Ross, R., JAOAC, 51, 1058-1062 (1968)
Captan and structurally related compounds: thin layer and gas-liquid chromatography
6, 7, 41

Pyysalo, H. et al., J. Chromatog., 168, 512-516 (1979)
Extraction and determination of o-phenylphenol and biphenyl in citrus fruits and apples
29, 56

Ragab, M.T.H. Anderson, M.G. & Johnston, H.W. Bull Envir. Contam. Toxicol. 44, 100-105 (1990)
Residue analysis of triadimefon, triadimenol and the BAY KWG1342 diol and BAY KWG1323
hydroxylated matabolites in winterweed.
133, 168.

Rains, D.M. & Holder, J.W., JAOAC, 64, 1252-1254 (1981)
Ethylene dibromide residues in biscuits and commercial flour
23

Rajzman, A., Analyst, 99, 120-127 (1974)
Determination of thiabendazole in citrus fruits by ultraviolet spectrophotometry
65

Reed, A.N., J. Chromatogr. 438, 393-400 (1988)
Quantification of triazole and pyrimidine plant growth retardants
161

Rosenberg, C. & Siltanen, H., Bull. Envir. Cont. Tox., 22, 475-478 (1979)
Residues of mancozeb and ethylenethiourea in grain samples
108

Roughan, J.A. et al., Analyst, 108, 742-747 (1983)
Modified gas-liquid chromatographic method for determining bromide/total bromine in foodstuffs and
soils
47

Sachse, J., Z. Lebensm. Unters. Forsch., 163, 274-277 (1977)
Über die Bestimmung von Chlorcholinchlorid (CCC) in Getreide
15

Sano, M. et al., JAOAC, 62, 764-768 (1979)
Flameless atomic absorption spectrophotometric determination of Vendex, an organic tin miticide, in
apples, oranges, and tea leaves
109

Saxton W.L et al. J. Agric. Food Chem., 37, 570-573 (1989)
Results of a survey for the presence of daminozide and unsymmetrical dimethylhydrazine in food.
104

Scudamore (1), K.A., Analyst, 105, 1171-1175 (1980)
Determination of 2-aminobutane in potatoes using high-performance liquid chromatography
89

Scudamore (2), K.A. & Goodship, G., Pest. Sci., 17, 385-395 (1986)
Determination of phosphine residues in fumigated cereals and other foodstuffs
46

Somerville, L., Meded. Fac. Landbouww. Rijksuniv. Gent, 45, 841-848 (1980)
The analysis of prochloraz residues in cereals
142

Specht (1), W. & Tillkes, M., Fres. Z. Anal. Chem., 307, 257-264 (1981)
Gas-chromatographische Bestimmung von Rückständen an Pflanzenbehandlungsmitteln nach Clean-up über Gel-Chromatographie und Mini-Kieselgel-Säulen-Chromatographie; 4. Mitteilung: Gas-chromatographische Bestimmung von 11 herbiciden Phenoxyalkancarbonsäuren und ihren Estern in Pflanzenmaterial
20, 121

Specht(2), W and Tilkes, M, Fres. Z. Anal. Chem., 322, 443-455 (1985)
Gas-chromatographische Bestimmung von Rückstanden von Pflanzenbehandlungsmitteln nach clean-up über Gel-chromatographie und Mini-Kieselgel-Säule Chromatographie, V. Methode zur aufarbeitung von Lebensmitteln und Futtermitteln plantzlicher und tierischer Herkunft für die bestimmung lipoid und wasserlöslicher Pflanzenbehandlungsmittel.
162

Stijve (1), T., Deutsche Lebensm. Rundsch., 77, 99-101 (1981)
Gas chromatographic determination of inorganic bromide residues - a simplified procedure
47

Stijve (2), T., IUPAC Pesticide Chemistry, Human Welfare and the Environment, J. Miyamoto (edit.), Pergamon Press, Oxford, UK, 95-100 (1983)
Miniaturised methods for monitoring organochlorine pesticide residues in milk
1, 21, 43, 44, 48, 71

Stijve (3), T. & Brand, E., Deutsche Lebensm. Rundsch., 73, 41-43 (1977)
A rapid, low cost, small-scale clean-up method for the determination of organochlorine pesticide residues in fats and oils
1, 12, 21, 43, 44, 48

Stijve (4), T., Deutsche Lebensm. Rundsch., 81, 321-322 (1985)
Inorganic bromide - a simple method for the confirmation of residue identity
47

Stijve (5), T., Deutsche Lebensm. Rundsch., 76, 234-237 (1980)
Thin-layer chromatographic determination of chlormequat residues in various substrates
15

Stijve (6), T., Deutsche Lebensm. Rundsch., 76, 119-122 (1980)
The determination of bromopropylate residues
70

Stijve (7), T., Challenges to Contemporary Dairy Analytical Techniques, Royal Society of Chemistry (London), Publ. no. 49, 293-302 (1984)
Determination and occurrence of organophosphorus pesticide residues in milk
4, 14, 17, 18, 19, 22, 25, 27, 28, 34, 37, 49, 60, 86, 90

Thornton, J.S., Olsen, T.J. and Wagber, K., Agr. Food Chem., 25, 573-576 (1977)
Determination of residues of metsystox-R and metabolite in Plant and animal tissue and soil
164

Tjan, G.H. & Konter, Th., JAOAC, 54, 1122-1123 (1971)
Gas-liquid chromatography of Morestan residues in plants
80

Tuinstra, L.G.M.Th. & Kienhuis, P.G.M., Chromatographia, 24, 696-700 (1987)
Automated two-dimensional HPLC residue procedure for glyphosate on cereals and vegetables with postcolumn fluoregenic labelling
158

VanHaver, W., Z. Lebensm. Unters. Forsch., 172, 1-3 (1981)
Determination of carbendazim and thiophanate-methyl residues in some vegetables and fruits by high pressure liquid chromatography
72, 77

VanWees, A.M.P. et al., in: Chromatography and Mass Spectrometry in Nutrition Science and Food Safety, A.Frigerio & H. Milon (edits.), Elsevier, Amsterdam, Netherlands, 19-25 A(1984)
Chromatographic methods for the determination of inorganic bromide in vegetables
47

Veierov, D. & Aharonson, N., JAOAC, 63, 532-535 (1980)
Economic method for analysis of fluid milk for organochlorine residues at the 10 ppb level
1 (not applicable to dieldrin), 12, 21, 43, 44, 48

Wagner, K. and Thornton, J.S. Pflanzensch. Nachr. Bayer, 30, 1-17 (1977)
Method for the gas-chromatographic determination of Metasystox(i)[R] and Metasystox R residues in plants, soil and water (German edit.: 30, 1-17 (1977).
73, 164, 166

Westcott. N.D., J Environm. Science and Health, 323, 317-330 (1988).
Terbufos residues in wheat and barley.
167

Wigfield, Y.Y. and Lanquette, M. JAOAC, 74. No.5 (1991)
Residue analysis of glyphosate and its principal metabolite in certain cereals, oilseeds and pulses by liquid chromatography and post-column fluorescence detection.
158

Winell, B., Analyst, <u>101</u>, 883-886 (1976)
Quantitative determination of ethoxyquin in apples by gas chromatography
35

Wright, D., JAOAC, <u>70</u>, 718-720 (1987)
New method for the determination of 1,1-dimethylhydrazine residues in apples and peaches
104

Yamada, T. et al., Agric. Biol. Chem., <u>48</u>, 1883-1885 (1984)
Determination of residual thiabendazole in citrus fruits and bananas by high performance liquid chromatography
65

Zimmerli, B. & Marek, B., Mitt. Geb. Lebensm. Unters. Hyg., <u>63</u>, 273-289 (1972)
Entwicklung einer gaschromatographischen Bestimmungs- und Bestätigungsmethode für Hexachlorbenzolrückstände in Fetten und Oelen
44

SECTION 5

DEFINITIONS AND REFERENCES

SECTION 5.1

DEFINITIONS FOR THE PURPOSE
OF THE CODEX ALIMENTARIUS

DEFINITIONS FOR THE PURPOSE
OF THE CODEX ALIMENTARIUS

The following definitions of terms are among those which have been adopted by the Codex Alimentarius Commission. These terms relate specifically to the topic of this Volume, pesticide residues. The list is not all inclusive.

Acceptable Daily Intake (ADI) of a chemical is the daily intake which, during an entire lifetime, appears to be without appreciable risk to the health of the consumer on the basis of all the known facts at the time of the evaluation of the chemical by the Joint FAO/WHO Meeting on Pesticide Residues. It is expressed in milligrams of the chemical per kilogram of body weight. (**Note:** For additional information on ADIs relative to pesticide residues refer to the Report of the 1975 Joint FAO/WHO Meeting on Pesticide Residues, FAO Plant Production and Protection Series No.1 or WHO Technical Report Series No. 592).

Animal Feed means harvested fodder crops, by-products of agricultural crops and other products of plant or animal origin which are used for animal feeding and which are not intended for human consumption.

Extraneous Residue Limit (ERL) refers to a pesticide residue or a contaminant arising from environmental sources (including former agricultural uses) other than the use of a pesticide or contaminant substance directly or indirectly on the commodity. It is the maximum concentration of a pesticide residue or contaminant that is recommended by the Codex Alimentarius Commission to be legally permitted or recognized as acceptable in or on a food, agricultural commodity, or animal feed. The concentration is expressed in milligrams of pesticide residue or contaminant per kilogram of the commodity.

Good Agricultural Practice in the Use of Pesticides (GAP) includes the nationally authorised safe uses of pesticides under actual conditions necessary for effective and reliable pest control. It encompasses a range of levels of pesticide applications up to the highest authorised use, applied in a manner which leaves a residue which is the smallest amount practicable. Authorised safe uses are determined at the national level and include nationally registered or recommended uses, which take into account public and occupational health and environmental safety considerations. Actual conditions include any stage in the production, storage, transport, distribution and processing of food commodities and animal feed.

Guideline Level is used to assist authorities in determining the maximum concentration of a pesticide residue resulting from a use reflecting good agricultural practice but where an acceptable daily intake or temporary acceptable daily intake for the pesticide has not been estimated or has been withdrawn by the Joint FAO/WHO Meeting on Pesticide Residues. The concentration is expressed in milligrams of pesticide residue per kilogram of the commodity. (**Note:** Guideline Levels are not to be advanced further than Step 4 in the Codex Procedure and are to be listed separately from MRLs and TMRLs in Codex documents).

Intake Study is a study designed to measure or estimate actual dietary exposures of consumers to pesticide residues or contaminants in order to compare such exposures to the acceptable daily intakes for pesticides or contaminants. (**Note:** For more information on intake studies, refer to Guidelines for the Study of Dietary Intakes of Chemical Contaminants prepared by the Joint FAO/WHO Food Contamination Monitoring Programme (WHO-EFP/83.53, FAO-ESN/MISC/83/2)).

Limit of Determination is the lowest concentration of a pesticide residue or contaminant that can be identified and quantitatively measured in a specified food, agricultural commodity, or animal feed with an acceptable degree of certainty by a regulatory method of analysis.

Maximum Residue Limit (MRL) is the maximum concentration of a pesticide residue (expressed as mg/kg), recommended by the Codex Alimentarius Commission to be legally permitted in or in food commodities and animal feeds. MRLs are based on GAP data and foods derived from commodities that comply with the respective MRLs are intended to be toxicologically acceptable. Codex MRLs which are primarily intended to apply in international trade, are derived from estimations made by the JMPR following:

a)   toxicological assessment of the pesticide and its residue; and

b)   review of residue data from supervised trials and supervised uses including those reflecting national food agricultural practices. Data from supervised trials conducted at the highest nationally recommended, authorised or registered uses are included in the review. In order to accommodate variations in national pest control requirements, Codex MRLs take into account the higher levels shown to arise in such supervised trials, which are considered to represent effective pest control practices.

Consideration of the various dietary residue estimates and determinations both at the national and international level in comparison with the ADI, should indicate that foods complying with Codex MRLS are safe for human consumption.

Pesticide means any substance intended for preventing, destroying, attracting, repelling, or controlling any pest including unwanted species of plants or animals during the production, storage, transport, distribution, and processing of food, agricultural commodities, or animal feeds or which may be administered to animals for the control of ectoparasites. The term includes substances intended for use as a plant-growth regulator, defoliant, desiccant, fruit thinning agent, or sprouting inhibitor and substances applied to crops either before or after harvest to protect the commodity from deterioration during storage and transport. The term normally excludes fertilizers, plant and animal nutrients, food additives and animal drugs. (**Note:** "Agricultural commodities" refers to commodities such as raw cereals, sugar beet, and cottonseed which might not, in the general sense, be considered food).

Pesticide Residue means any specified substances in food, agricultural commodities, or animal feed resulting from the use of a pesticide. The term includes any derivatives of a pesticide, such as conversion products, metabolites, reaction products, and impurities considered to be of toxicological significance. (**Note:** The term "pesticide residue" includes residues from unknown or unavoidable sources (e.g., environmental), as well as known uses of the chemical).

Regulatory Method of Analysis is a method that has been validated and can be applied using normal laboratory equipment and instrumentation to detect and determine the concentration of a pesticide residue or contaminant in a food, agricultural commodity, or animal feed for purposes of determining compliance with a maximum residue limit or extraneous residue limit. (**Note:** For more information on regulatory methods of analysis and their application, refer to Recommendation for Methods of Analysis for Pesticide Residues and Codex Guidelines on Good Analytical Practice (see Section 4 of this Volume, parts 4.2 and 4.3).

Temporary Acceptable Daily Intake (TADI) is an acceptable daily intake established for a specified, limited period to enable additional biochemical, toxicological or other data to be obtained as may be required for estimating an acceptable daily intake. (**Note:** A TADI estimated by the Joint FAO/WHO Meeting on Pesticide Residues normally involves the application of a safety factor larger than that used in estimating an ADI).

Temporary MRL (TMRL) or Temporary ERL (TERL) is an MRL or ERL established for a specified, limited period and is recommended under either of the following conditions:

1. Where a temporary acceptable daily intake has been estimated by the Joint FAO/WHO Meeting on Pesticide residues for the pesticide or contaminant of concern; or

2. Where, although an acceptable daily intake has been estimated, the good agricultural practice is not sufficiently known or residue data are inadequate for proposing an MRL or ERL by the Joint FAO/WHO Meeting on Pesticide Residues.

(**Note:** TMRLs and TERLs are not to be advanced further than Step 7 of the Codex Procedure).

SECTION 5.2

REFERENCES TO
PREVIOUS CODEX PUBLICATIONS
ON CODEX MRLS

REFERENCES TO
PREVIOUS CODEX PUBLICATIONS
ON CODEX MRLS

1. Codex Maximum Limits for Pesticide Residues adopted by the Codex Alimentarius Commission up to the end of the 16th Session, July 1985. CAC/VOL. XIII - Ed. 2 (1986).

2. Codex MRLs and Amendments adopted at the 17th Session of the Codex Alimentarius Commission. Supplement 1 to CAC/VOL. XIII - Ed. 2 (1988).

3. Codex MRLs and Amendments adopted at the 18th Session of the Codex Alimentarius Commission. Supplement 2 to CAC/VOL. XIII - Ed. 2 (1989).

4. Guide to Codex Maximum Limits for Pesticide Residues:

    Part 1 - General Notes and Guidelines. CAC/PR 1-1984.

    Part 2 - Maximum Limits for Pesticide Residues. CAC/PR 2. Discontinued.[1]

    Part 3 - Guideline Levels for Pesticide Residues. CAC/PR 3. Discontinued.[1]

    Part 4 - Classification of Foods and Animal Feeds. CAC/PR 4-1989 (English only).

    Part 5 - Recommended Method of Sampling for Determination of Pesticide Residues. CAC/PR 5-1984.

    Part 6 - Portion of Commodities to which Codex Maximum Residue Limits Apply and which is Analysed. CAC/PR 6-1984.

    Part 7 - Codex Guidelines on Good Practice in Pesticide Residues Analysis. CAC/PR 7-1984.

    Part 8 - Recommendations for Methods of Analysis of Pesticide Residues. CAC/PR 8-1989, Fourth Edition.

    Part 9 - Recommended National Regulatory Practices to Facilitate Acceptance and Use of Codex Maximum Limits for Pesticide Residues in Foods.

---

[1] Publication as Codex Final Texts has been discontinued. Most of the contents of CAC/PR 2 are contained in CAC/VOL. XIII - Ed. 2 and supplements in occasional working papers.

SECTION 5.3

REFERENCES TO
FAO/WHO PUBLICATIONS
RELATED TO MRLS

**REPORTS AND OTHER DOCUMENTS RESULTING FROM PREVIOUS JOINT
MEETINGS OF THE FAO PANEL OF EXPERTS ON PESTICIDE RESIDUES
IN FOOD AND THE ENVIRONMENT AND WHO EXPERT GROUPS
ON PESTICIDE RESIDUES**

PREVIOUS FAO AND WHO DOCUMENTS

1. Principles governing consumer safety in relation to pesticide residues. Report of a meeting of
    a WHO Expert Committee on Pesticide Residues held jointly with the FAO Panel of Experts
    on the Use of Pesticides in Agriculture. FAO Plant Production and Protection Division Report,
    No. PL/1961/11; WHO Technical Report Series, No. 240.

2. Evaluation of the toxicity of pesticide residues in food; Report of a Joint Meeting of the FAO
    Committee on Pesticides in Agriculture and the WHO Expert Committee on Pesticide Residues.
    FAO Meeting Report, No. PL/1963/13; WHO/Food Add./23.

3. Evaluation of the toxicity of pesticide residues in food. Report of the Second Joint Meeting of
    the FAO Committee on Pesticides in Agriculture and the WHO Expert Committee on Pesticide
    Residues. FAO Meeting Report, No. PL/1965/10; WHO/Food Add./26.65.

4. Evaluation of the toxicity of pesticide residues in food. FAO Meeting Report, No.
    PL/1965/10/1; WHO/Food Add./27.65.

5. Evaluation of the hazards to consumers resulting from the use of fumigants in the protection of
    food. FAO Meeting Report, No. PL/1965/10/2; WHO/Food Add./28.65.

6. Pesticide residues in food. Joint report of the FAO Working Party on Pesticide Residues and
    the WHO Expert Committee on Pesticide Residues. FAO Agricultural Studies, No. 73, WHO
    Technical Report Series, No. 370.

7. Evaluation of some pesticide residues in food. FAO/PL:CP/15; WHO/Food Add./67.32.

8. Pesticide residues. Report of the 1967 Joint Meeting of the FAO Working Party and the WHO
    Expert Committee. FAO Meeting Report, No. PL:1967/M/11; WHO Technical Report Series,
    No. 391.

9. 1967 Evaluations of some pesticide residues in food. FAO/PL:1967/M/11/1; WHO/Food
    Add./68.30.

10. Pesticide residues in food. Report of the 1968 Joint Meeting of the FAO Working Party of
    Experts on Pesticide Residues and the WHO Expert Committee on Pesticide Residues. FAO
    Agricultural Studies, No. 78; WHO Technical Committee Report Series, No. 417.

11. 1968 Evaluation of some pesticide residues in food. FAO/PL:1968/M/9/1; WHO/Food
    Add./69.35.

12. Pesticide residues in food. Report of the 1969 Joint Meeting of the FAO Working Party of Experts on Pesticide Residues and the WHO Expert Group on Pesticide Residues. FAO Agricultural Studies, No. 84; WHO Technical Report Series, No. 458.

13. 1969 Evaluations of some pesticide residues in food. FAO/PL:1969/M/17/1; WHO/Food Add./70.38.

14. Pesticide residues in food. Report of the 1970 Joint Meeting of the FAO Working Party of Experts on Pesticide Residues and the WHO Expert Committee on Pesticide Residues. FAO Agricultural Studies, No. 87; WHO Technical Report Series, No. 474.

15. 1970 Evaluations of some pesticide residues in food. AGP:1970/M/12/1; WHO/Food Add./71.42.

16. Pesticide residues in food. Report of the 1971 Joint Meeting of the FAO Working Party of Experts on Pesticide Residues and the WHO Expert Committee on Pesticide Residues. FAO Agricultural Studies, No. 88; WHO Technical Report Series, No. 502.

17. 1971 Evaluations of some pesticide residues in food. AGP-1971/M/9/1; WHO Pesticide Residues Series, No. 1.

18. Pesticide residues in food. Report of the 1972 Joint Meeting of the FAO Working Party of Experts on Pesticide Residues and the WHO Expert Committee on Pesticide Residues. FAO Agricultural Studies, No. 90; WHO Technical Report Series, No. 525.

19. 1972 Evaluations of some pesticide residues in food. AGP-1972/M/9/1; WHO Pesticide Residues Series, No. 2.

20. Pesticide residues in food. Report of the 1973 Joint Meeting of the FAO Working Party of Experts on Pesticide Residues and the WHO Expert Committee on Pesticide Residues. FAO Agricultural Studies, No. 92; WHO Technical Report Series, No. 545.

21. 1973 Evaluations of some pesticide residues in food. FAO/AGP/1973/M/9/1; WHO Pesticide Residues Series, No. 3.

22. Pesticide residues in food. Report of the 1974 Joint Meeting of the FAO Working Party of Experts on Pesticide Residues and the WHO Expert Committee on Pesticide Residues. FAO Agricultural Studies, No. 97; WHO Technical Report Series, No. 574.

23. 1974 Evaluations of some pesticide residues in food. FAO/AGP/1974/M/11; WHO Pesticide Residues Series, No. 4.

24. Pesticide residues in food. Report of the 1975 Joint Meeting of the FAO Working Party of Experts on Pesticide Residues and the WHO Expert Committee on Pesticide Residues. FAO Plant Production and Protection Series No. 1; WHO Technical Report Series, No. 592.

25. 1975 Evaluations of some pesticide residues in food. AGP:1975/M/13; WHO Pesticide Residues Series, No. 5.

26. Pesticide residues in food. Report of the 1976 Joint Meeting of the FAO Panel of Experts on Pesticide Residues and the Environment and the WHO Expert Group on Pesticide Residues. FAO Food and Nutrition Series No. 9; FAO Plant Production and Protection Series No. 8; WHO Technical Report Series, No. 612.

27. 1976 Evaluations of some pesticide residues in food. AGP:1976/M/14.

28. Pesticide residues in food - 1977. Report of the Joint Meeting of the FAO Panel of Experts on Pesticide Residues and the Environment and the WHO Expert Group on Pesticide Residues. FAO Plant Production and Protection Paper 10, Rev.

29. Pesticide residues in food: 1977 Evaluations. FAO Plant Production and Protection Paper 10, Sup.

30. Pesticide residues in food - 1978. Report of the Joint Meeting of the FAO Panel of Experts on Pesticide Residues and the Environment and the WHO Expert Group on Pesticide Residues. FAO Plant Production and Protection Paper 15.

31. Pesticide residues in food: 1978 Evaluations. FAO Plant Production and Protection Paper 15, Sup.

32. Pesticide residues in food - 1979. Report of the Joint Meeting of the FAO Panel of Experts on Pesticide Residues and the Environment and the WHO Expert Group on Pesticide Residues. FAO Plant Production and Protection Paper 20.

33. Pesticide residues in food: 1979 Evaluations. FAO Plant Production and Protection Paper 20, Sup.

34. Pesticide residues in food - 1980. Report of the Joint Meeting of the FAO Panel of Experts on Pesticide Residues and the Environment and the WHO Expert Group on Pesticide Residues. FAO Plant Production and Protection Paper 26.

35. Pesticide residues in food: 1980 Evaluations. FAO Plant Production and Protection Paper 26, Sup.

36. Pesticide residues in food - 1981. Report of the Joint Meeting of the FAO Panel of Experts on Pesticide Residues and the Environment and the WHO Expert Group on Pesticide Residues. FAO Plant Production and Protection Paper 37.

37. Pesticide residues in food: 1981 Evaluations. FAO Plant Production and Protection Paper 42.

38.   Pesticide residues in food - 1982. Report of the Joint Meeting of the FAO Panel of Experts on Pesticide Residues and the Environment and the WHO Expert Group on Pesticide Residues. FAO Plant Production and Protection Paper 46.

39.   Pesticide residues in food: 1982 Evaluations. FAO Plant Production and Protection Paper 49.

40.   Pesticide residues in food - 1983. Report of the Joint Meeting of the FAO Panel of Experts on Pesticide Residues and the Environment and the WHO Expert Group on Pesticide Residues. FAO Plant Production and Protection Paper 56.

41.   Pesticide residues in food: 1983 Evaluations. FAO Plant Production and Protection Paper 61.

42.   Pesticide residues in food - 1984. Report of the Joint Meeting on Pesticide Residues. FAO Plant Production and Protection Paper 62.

43.   Pesticide residues in food: 1984 Evaluations. FAO Plant Production and Protection Paper 67.

44.   Pesticide residues in food - 1985. Report of the Joint Meeting of the FAO Panel of Experts on Pesticide Residues and the Environment and a WHO Expert Group on Pesticide Residues. FAO Plant Production and Protection Paper 68.

45.   Pesticide residues in food: 1985 Evaluations. Part I - Residues. FAO Plant Production and Protection Paper 72/1.

46.   Pesticide residues in food: 1985 Evaluations. Part II - Toxicology. FAO Plant Production and Protection Paper 72/2.

47.   Pesticide residues in food - 1986. Report of the Joint Meeting of the FAO Panel of Experts on Pesticide Residues and the Environment and a WHO Expert Group on Pesticide Residues. FAO Plant Production and Protection Paper 77.

48.   Pesticide residues in food: 1986 Evaluations. Part I - Residues. FAO Plant Production and Protection Paper 78.

49.   Pesticide residues in food: 1986 Evaluations. Part II - Toxicology. FAO Plant Production and Protection Paper 78/2.

50.   Pesticide residues in food - 1987. Report of the Joint Meeting of the FAO Panel of Experts on Pesticide Residues and the Environment and a WHO Expert Group on Pesticide Residues. FAO Plant Production and Protection Paper 84.

51.   Pesticide residues in food: 1987 Evaluations. Part I - Residues. FAO Plant Production and Protection Paper 86/1.

52.   Pesticide residues in food: 1987 Evaluations. Part II - Toxicology. FAO Plant Production and Protection Paper 86/2.

53.   Pesticide residues in food - 1988. Report of the Joint Meeting of the FAO Panel of Experts on Pesticide Residues and the Environment and a WHO Expert Group on Pesticide Residues. FAO Plant Production and Protection Paper 92.

54.   Pesticide residues in food: 1988 Evaluations. Part I - Residues. FAO Plant Production and Protection Paper 93/1.

55.   Pesticide residues in food: 1988 Evaluations. Part II - Toxicology. FAO Plant Production and Protection Paper 93/2.

56.   Pesticide residues in food - 1989. Report of the Joint Meeting of the FAO Panel of Experts on Pesticide Residues and the Environment and a WHO Expert Group on Pesticide Residues. FAO Plant Production and Protection Paper 99.

57.   Pesticide residues in food: 1989 Evaluations. Part I - Residues. FAO Plant Production and Protection Paper 100.

58.   Pesticide residues in food: 1989 Evaluations. Part II - Toxicology. FAO Plant Production and Protection Paper 100/2.

59.   Pesticide residues in food - 1990. Report of the Joint Meeting of the FAO Panel of Experts on Pesticide Residues and the Environment and a WHO Expert Group on Pesticide Residues. FAO Plant Production and Protection Paper 103.

60.   Pesticide residues in food: 1990 Evaluations. Part I - Residues. FAO Plant Production and Protection Paper 103/1.

61.   Pesticide residues in food: 1990 Evaluations. Toxicology. World Health Organization, WHO/PCS/91.47.

62.   Pesticide residues in food - 1991. Report of the Joint Meeting of the FAO Panel of Experts on Pesticide Residues and the Environment and a WHO Expert Group on Pesticide Residues. FAO Plant Production and Protection Paper 111.

63.   Pesticide residues in food: 1991 Evaluations. Part I - Residues. FAO Plant Production and Protection Paper 113/1.

64.   Pesticide residues in food: 1991 Evaluations. Part II - Toxicology. FAO Plant Production and Protection Paper 113/2.